2-3-10

For the Love of Belle

Cushenbury—Big Bear

Walter Del Mar and Martha Owens

WALTER DEL MAR AND MARTHA OWENS

FOR
THE
LOVE
OF

Belle

A MINER STRIKES GOLD

CUSHENBURY—BIG BEAR

Epic
Press

Belleville, Ontario, Canada

FOR THE LOVE OF BELLE
Copyright © 2010, Walter Del Mar & Martha Owens

Library and Archives Canada Cataloguing in Publication

Del Mar, Algernon, b. 1870
     For the love of Belle / compiled and edited by Walter
Del Mar and Martha Owens.

A collection of letters exchanged between Algernon Del Mar
     and Belle Rogers between 1900 to 1903.
ISBN 978-1-55452-455-6

     1. Del Mar, Algernon, b. 1870--Correspondence.
2. Rogers, Belle--Correspondence. 3. Mining
engineers--California--Correspondence.
4. California--Biography. I. Del Mar, Walter, 1920-
II. Owens, Martha, 1937- III. Rogers, Belle IV. Title.

TN140.D44A3 2009   622.092   C2009-907101-0

To order additional copies, visit:
www.essencebookstore.com

For more information, please contact:
Walter Del Mar
44-816 Oro Grande Circle
Indian Wells, CA 92210

Epic Press is an imprint of *Essence Publishing,* a Christian Book Publisher dedicated to furthering the work of Christ through the written word. For more information, contact:
20 Hanna Court, Belleville, Ontario, Canada K8P 5J2
Phone: 1-800-238-6376 • Fax: (613) 962-3055
E-mail: info@essence-publishing.com
Web site: www.essence-publishing.com

Introduction

In 1998 Walter Del Mar wrote *The Gold Mines of Black Hawk Canyon* as a way to preserve both the photos and the history of the mine that had played such an important part in the life of his family. This current book is also about mining and the Del Mar family, but with a twist.

After his mother died in 1962, Walter Del Mar found a box of letters among her things. They were the letters that Algernon Del Mar, Walter's father, had written to his future wife, Belle Rogers, during the time of their courtship and engagement. The letters cover the years 1900 to 1903.

While we originally discussed making the letters into a novel, we decided that the letters tell such a wonderful story by themselves that doing a little editing and adding few notes of explanation was perhaps the better way to present this "slice of life" glimpse into an earlier time.

The letters show how Del Mar and his bride waited three years to get married. Del Mar insisted he have a job that produced enough income for him to support a wife and, eventually, a family. During the period of the engagement, Del Mar tried always to have a position in a location not terribly distant from Belle's family home in Pasadena.

The reader will quickly realize that material printed in italics is original text from the letters. Notes and comments we have added are in square brackets. Quotations from Algernon Del Mar's memoirs or work journal (written after he retired) are in regular print and are acknowledged in the text. Much of the information about mining techniques and terms comes from Walter Del Mar and is based on his direct experience with mines.

One further note about the text: in some instances the spelling of places is not consistent. For example, sometimes Del Mar wrote "Black Hawk" and other times it was "Blackhawk," a single word. The same is true in speaking of another mine, the Bluebird, some-times the Blue Bird. We have left the spellings as they are in the original. The name San Burdoo occurs several times in Del Mar's letters. San Burdoo is a local short name for San Bernardino.

This is a short read, but a good read. All the elements of a good drama are here as the story of the young couple unfolds.

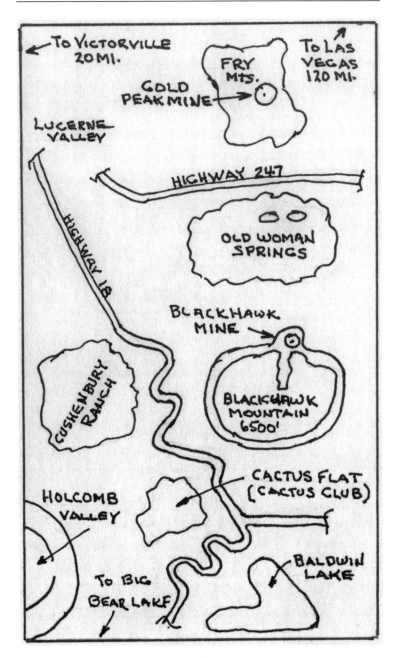

The Beginning:
A Miner's Love Story

This is a story about mining in California and about gold mining in particular. It is the story about two people in love and the struggle to make a good life in a period when income from mining was erratic. This story is set in a time and place when it was often a struggle to have a family life and being a miner's wife often meant a rugged and unpredictable life.

Algernon Del Mar lived much of his early life with his family in San Francisco and learned about gold mining first from his father, Alexander Del Mar. After completing his studies in secondary school, he planned to proceed on to college. Mining as a curriculum was new to the United States in the 1880s, and most mining colleges were in Europe—primarily in Germany, Spain, and England. Del Mar was accepted to the Royal School of Mines in London and graduated with honors in 1891. With his diploma under his arm, he was ready to make his mark in the real world. A mining engineer was much in demand as the world was alive with dreams of getting rich from the earth's unknown wealth.

Del Mar was off first to South Africa—quite a trip by ship—and into the interior to report on the mines owned by rich English men. Following this assignment, he went to Norway before returning to the United States, where he was hired to report on some coal mines in the southeast.

It was natural at this time in Del Mar's life to think about the place where he first did some mining work with his father. Accordingly, he left for the San Bernardino Mountains near Big Bear Valley. The mine he sought was in Holcomb Valley, an area where there was a great find of gold in the 1860s. Holcomb Valley is high in the mountains and was not a pleasant place to be in the winter months. Young Del Mar worked his claim here in the high mountains while traveling back and forth from the desert valley. He had decided soon after graduation that he would not work for a large company. He wanted to make his mark as an entrepreneur.

George Rogers, father of Belle, came to California as a child. His family had come from the Boone, KY, area and traveled to Peoria, IL, where they joined other families for the long trek west. Once in California the Rogers group had traveled from Bodie, to Lompoc, then to Santa Barbara, next to West Los Angeles, and, finally, to Pasadena, where they settled.

At the time this story begins travel by horse and buggy from Pasadena to the mountains was a week long trek through Cajon Pass and across the high desert to the mountain pass leading to Big Bear Lake. Del Mar family lore says that Algernon Del Mar and Belle Rogers met at the area known as Cushenbury Ranch, a natural oasis with a fresh water spring that is located in the

Lucerne Valley of California's high desert area. The exact date is not recorded in any family history, but, as the letters that follow will show, they apparently met while Belle was still in high school. (She graduated from Pasadena High School in 1900.) Del Mar, twelve years her senior, worked then in the general area of Big Bear, particularly the Black Hawk Mountain, a mine his father had helped develop earlier and to which the son had now returned.

The trip from the Rogers' home in Pasadena and across the desert to the area of Big Bear Lake in the San Bernardino Mountains meant staying over several nights on the road. Their route led them from Pasadena through Monrovia, Arcadia, San Bernardino, Cajon Pass, and across the high desert to Lucerne Valley. In the 1890s the overnight stays were frequently done at stage stops and almost always meant camping out. It was a trip the George Rogers' family had made often, generally while on the way from Pasadena to cooler summer weather in the area of Big Bear Lake. They traveled by horse and wagon and took with them a tent, bedding, cooking and eating utensils, and a few chairs. Often it was Rosa (Ma) Rogers who drove the family there as George Rogers, her husband, was frequently on the road with his business endeavors.

Along this route into the mountains, Cushenbury Ranch was a natural oasis located in the north foothills of the San Bernardino Mountains. Many shade trees offered a cool and protected camping area. Natural springs turned the nearly 120 acre area into a favorite stopping point for travelers. At one time it was also a stage stop as well as a watering hole for all travelers who were headed up the steep grade to Bear Valley and Holcomb Valley.

Frequent travelers, such as the Rogers family, became acquainted with others, including the miners and prospectors who traveled the area and who frequented this spot. Del Mar and the entire Rogers family encountered each other on several trips. George Rogers, Belle's father, was a mule team contractor and a horse dealer who apparently met Del Mar on occasion when his teams were used at various mines. Rogers was also a contract mail carrier who traveled out from Victor (now Victorville) to the small settlements in the area. One can suppose that Del Mar and Rogers met each other in these camping spots more than once and that Rogers then introduced Del Mar to his family.

George and Rosa Rogers had four children at the time Del Mar met them. (A fifth child had died in infancy.) The oldest child, named Adella Belle but called Belle (born 1882), is a main character in this narrative. She had younger sisters Lotus (born 1883) and Reine (born 1889) and a brother Bruce (born 1897). Belle and Lotos were especially close, not just in age, but as best friends as well as sisters. They lived in a typical small wooden frame house in Pasadena near what is now Orange Grove Ave. and Colorado Blvd. Streets were not paved, and the large yards along the street had many trees but little grass. Animals, including horses, were kept in the barn or carriage house on the property.

By the time Belle was a senior at Pasadena High School in 1899 to 1900, it was obvious that there was a strong attraction between her and Del Mar. Pictures show that at the time this story begins, Belle was a very attractive dark-haired young woman. The fact that she rode in the Tournament of Roses Parade in 1900, long before the elaborate floats of today, attests to her good looks.

Del Mar, a New Yorker by birth, grew up in San Francisco. His father, Alexander Del Mar, was a mining engineer. By the time Del Mar and Belle met, Del Mar already had a reputation for his expertise in metallurgy and mining after assignments in South Africa, Norway, Canada, Mexico, and the United States. Love would soon determine that his assignments be closer to home.

Mining engineering in the early 1900s was not a stable career. It was a great struggle to establish steady income as many mines were short-lived with ore bodies that failed to produce to expectations. Many people expected to get rich with gold. A few did; most did not. In his own memoirs, Del Mar recalls that his father always told him, "As a rule the sale of a going mine is usually worth the sum of two and can never be more than the sum of four years profit." The reason for the statement was, of course, to highlight the short life of a mine. The story that follows will show how often this short life is true.

Life in mining camps was not easy. Housing was tents and cabins in most places. Food and water supplies were limited. One had to battle the elements—desert heat and winter snows—without the benefit of our modern climate control technology. Communication was primarily letters and word of mouth. Telegraph offices and telephones existed, yes, but were normally many miles from the mine. The work that today is handled by machines was largely done then by strong arms and backs.

By 1900 Del Mar was thirty and had returned to the Big Bear mining area that he knew as a teenager. His father, Alexander, had worked and raised money for the development of mining in the Holcomb Valley of the San Bernardino Mountains. Del Mar was familiar with the area and apparently was among those who believed there was still gold to be taken from this place. In particular, he was acquainted with the Black Hawk Mountain mining area that had been partially owned by his father at one time.

What follows are excerpts from letters written by Del Mar beginning first before he and Belle were actually engaged and then covering the entire period of their extended courtship.

The first letter, dated December 5, 1899, is written from the Black Hawk Mine. His letters to Belle at this time begin with the very proper greeting of *My Dear Miss Belle*. His handwriting is the elaborate and handsome handwriting of past generations.

One further note—in the letters and elsewhere the name of the mine is usually written as two words, *Black Hawk*, but occasionally it is found as a single word, *Blackhawk*. The spellings used follow that of the original letters.

December 5, 1899 Black Hawk Mines

"My Dear Miss Belle"

In this earliest letter Del Mar thanks Belle for the recipe (called *receipt* then) for corn cake she has sent and also for the violets that Reine has sent—their smell, he writes, is a pleasant change from sage.

The recipe was of importance, for apparently he has tried before to make corn cake (cornbread) in the rather primitive cooking facilities of a mining camp. In recounting his attempts to make corn cake, he writes, *"What puzzled me was to get the cake to remain whole and not crumble. I see now that eggs and molasses will do the trick. The first time I took the pan out of the oven [was] to turn the cake over, as our oven cooks imperfectly on the bottom. In the act of turning the cake it went to pieces but we ate it just the same and found it good. The second time I distinguished myself and we had the best corn cake I have ever eaten."*

Apparently when she sent the recipe, Belle referred to Del Mar as "cook," and Del Mar tells her she is lucky she wrote *"as you are cook"* in brackets or he would have *"vented my wrath in unmistakable terms."* He is reminded of how when he was at home, he was explaining to some friends how he had to teach the natives in South Africa to cook. *"The parlor maid overheard part of the conversation and as usual with servants she retailed* [does he mean *relayed?*] *the information that I was a cook in a mining camp to several of her friends. A short while later, a friend told his mother that her maid told her he was a mining camp cook."* The family got a good laugh from the incident.

Weather in the high desert of Black Hawk Mine was very different from that in Pasadena. Winters in the mountain mine area tend to be cold and very windy. Del Mar asks Belle not to send him weather news or sympathy for his weather. In the three previous months, he reports, they had about five days with no snow at all. Twice there was snow at the Santa Fe (a reference to the top of Black Hawk Mountain), but it melted the following day. *"We are having fine clear bracing weather, in fact ideal weather and we are working like trojans."*

Throughout the letters, Del Mar recounts facts and stories involving residents—often miners—of the area. One man who appears at various times is MacFee, referred to here as *Mac*. In more than one place he is called *Mr. J. E. MacFee,* and he lived in or near the settlement called Cactus Flat located at the top of Cushenbury Canyon. (The Cactus Flat area is now accessed off SR 18.)

In this early letter, Del Mar reports, *"By the way, Mac has shaved off his mustache. Poor fellow. He hates grey hairs so to save the trouble of pulling them out singly he has butchered the whole lot. He now looks like Billy Bryan out of a job or a pilot who has lost his bearing."* (Bill Bryan may be a reference to Wm. Jennings Bryan.)

The following excerpt from the letter presents a bit of a puzzle. Del Mar reports that he and Mac worked on Thanksgiving Day since there was nothing else to do. *"As Cushenbury already has a record of one baked cat [?] I did not go there Thanksgiving for fear of another tragedy. One can never tell what will happen in the 'spiritual' atmosphere."* Del Mar's reference to a "spiritual atmosphere" means an atmosphere dominated by drinking spirits. The reader can draw his own conclusion regarding the baked cat.

Other people mentioned include a Mrs. Matheson, whose position in the area is unknown.

"You probably know how I loathe the sight of Mrs. Matheson," Del Mar writes. *"Well, just imagine her in black stockings with petticoats to the knees and you can appreciate my feelings last Wednesday at Cactus Flats. It took all my self control to suppress my laughing in her face. My sense of the ridiculous received its Waterloo. She asked me very cordially to pay them a visit which invitation I cordially declined. The she had the cheek to say to Mr. Johnston that 'Mr. Del Mar will try to get up to see us when Mrs. Foltz comes.' I guess not. I value my good reputation too highly to mix with such people."*

On the domestic front, Del Mar reports his chickens are so tame they follow him and jump on his lap. He says his chickens get four eggs a day while Johnston has three times as many chickens and gets no eggs. *"I have learned that during moulting season the hens must be well fed so I fill up Billy and his wives until they get tired of eating. By this means we have had eggs right along for while one is laying another is moulting."*

17

Another person who enters the letters at various times is a man known as Cactus Flat Jim Johnston. Del Mar tells us in his memoirs that Johnston *"kept an eating station of sorts at Cactus Flat."*

In his work diary Del Mar adds that Johnston was a character. "He cultivated some fruit trees but the climate, I guess, was too cold for they did not do very well." Del Mar adds that he was always welcome at Johnston's place.

"The stage to the Rose Mine always stopped there [at Johnston's], sometimes for lunch or to give the horses a rest after the long pull up Cushenbury Canyon. Often the stage stopped at Cushenbury Ranch at the foot of the grade where water was handy from a water trough at the side of the road. Either MacFee or Mr. Leach was there to greet the stage passengers."

Del Mar also notes in this same diary that his father paid the expense of keeping either MacFee or Leach at the ranch and later his brother Eugene Del Mar gave them money to do assessment work. "Leach had a horse,' Del Mar writes in his diary, "which I often rode to the mine after I had walked to Cushenbury. I simply let the horse loose and he would find his way back to the ranch."

Del Mar was an avid photographer and left many pictures of the places he lived and worked. In this first letter he tells of trying to get a picture of Billy Barnacle (the rooster) and the cat. The result was three ruined films *"because the cat was too suspicious."*

Del Mar writes here that he expects to go to town on the 21st and says he will contact Belle's mother about a date on which he can visit the family.

Christmas 1899 brought young Del Mar to Pasadena for several days, during which he took many pictures of the young woman he now loved. The resulting photos, however, were not up to the quality he wanted.

1900

January 9, 1900
Blackhawk

Del Mar writes that the new toning solution he used on the photos might be to blame. He was particularly critical of pictures he took of Lotus, Belle's younger sister. He writes that the picture *"outside your house somehow does not look like her and is a poor negative."*

It is interesting to note that the letter, written in his fine, almost spidery penmanship, begins *"My dear Miss Belle"* and the word "Miss" has the first "s" resembling the letter "f" as was once the style for writing a double "s."

After an all too short trip to Pasadena, Del Mar returns to the Black Hawk. Although Del Mar frequently used Victor as a spot where he was or where his mail was to be directed, the town he called Victor is now Victorville. The name was changed early in the 20th century because two California towns with the name "Victor" existed. Black Hawk mine on Blackhawk Mountain is thirty miles east of the current Victorville.

Victor was an important town because it was a railroad stop as well as a stage stop, and it had stores for supplies for the mines and miners.

January 9, 1900
Black Hawk, Victor, California

Please notice that the address above is *Victor*. The Black Hawk Mine is not located at Victor but, here and throughout the letters, a town listed by the letter's date is the post office to which a reply can be sent. There was seldom a post office at a mine camp.

Life at the mine was not without problems. Travel problems did not help. After spending an extra day on the road because the train to Victor arrived too late for Del Mar to make the needed connection to the stage going to the Black Hawk, he found his friend and co-worker Leach had not recovered from welcoming in the New Year. Apparently looking for better company, Del Mar went to Cactus Flats, several miles away, in search of more compatible company. He didn't find it, as he writes, *"Johnston was out of temper as usual and I was glad to get the mule...over the hills to my farmyard and our friend MacFee."* Here both Billie Barnacle (the rooster) and a cat soundly welcomed him. MacFee, he noted, was in good spirits, unlike Johnston and Leach.

Blackhawk Canyon is a narrow gorge running in a northerly direction to the Mojave Desert. In the upper part of the canyon, the tents of the camp were located near a spring. It was not an easy life. Gold mining was different from placer mining, which looked for gold in streams. Mining at Black Hawk and other spots meant tunneling into the side of a mountain. Unlike other types of mining, such as coal or copper mining, the shafts generally did not descend down into the earth but stayed relatively level with the point of entry into the mountain.

Additional shafts might radiate from the original tunnel into the mountain, and each shaft was named. A tunnel and its shafts needed to be kept in constant repair for the safety of the miners who worked there. One tunnel mentioned at the Blackhawk was the Santa Fe.

January in these mountains sometimes brings brief intervals of snow, but the weather generally stays a bit above freezing. Inside the mineshafts temperatures remain cool but constant and work could continue. A vein of ore did not necessarily remain in a straight line. Del Mar writes of building a raise, which is an incline that allows the miner to access a vein that has taken an upward turn. However, this particular incline gave him some problems.

Building the incline meant looking up as Del Mar used a pick to remove rock and fine dirt. His natural good humor made light of the uncomfortable job.

"It seems," he wrote,*" that every piece of dirt that flies off gets lodged in my right eye so that I hope soon to have a potato patch growing on the accumulated soil."*

The letters often contain references to people who apparently lived in the general area of the mines, most often those in the Blackhawk area. In early 1900 Del Mar speaks of Mrs. Madison as having another fish (meaning of *fish* unclear here). The "fish" is supposedly a rich widow from Kansas whom Johnston claims he will marry if she is really rich. Del Mar has doubts about the "fish" and thinks she will *"turn out to be a raw boned spiritualist with a small purse so will not bother about making her acquaintance, especially as birds of a feather often flock with one another. She intends to stay in Holcomb all winter."*

February 27, 1900
Black Hawk, Victor, California

Del Mar still addresses Belle as *"My Dear Miss Belle"* but in the early part of the letter notes that since she has just celebrated her eighteenth birthday, *"I think I may take the liberty of addressing you direct instead of through your mother."*

A softer side of Del Mar shows as he thanks her for the book and photograph she has sent. *"I am not generally considered sentimental,"* he writes, *"but I think I am in respect to photographs for the pleasure I take in looking at them can only be appreciated by those who live away from home and friends."*

MacFee has been gone for a month, and Del Mar notes he doesn't where he has gone or if he is even still alive. As for work, Del Mar has left the Blackhawk and is now running the mill and cyanide plant at Bear Valley for Eberle Brothers.

In Del Mar's own writing of his work history he explains that the Eberle brothers "came to me with the story that they had found rich float [surface ore] near the Gold Mountain Mine in Bear Valley. They wanted me to save the gold for them. They provided two pairs of Cornish Rolls [devices for crushing the ore] and a copper plate. I did not take to the idea. I knew that rolls would not grind fine enough to free the gold but I determined to try."

A cyanide plant, such as Del Mar was now running, was a gold recovery system that used cyanide to extract the gold from the ore. This method supplemented the earlier practice of using only mercury to recover the gold. Before the cyanide could be used, the ore had to be screened and then crushed. In 1900, a stamp mill would have been used for the crushing.

After being crushed, the ore was washed over plates coated with mercury to pick up the free gold (i.e., gold now free of surrounding rock). Next the ore was treated with a cyanide solution. This was normally done in a large tank.

Up to a day later, the solution containing the cyanide and the gold was drained off. The rest of the ore went to the tailings as waste. The gold-bearing cyanide solution was treated in tanks of loose zinc, which collected the gold and allowed the liquid to pass as waste. Next the zinc and gold mixture was fired in a furnace, where the zinc was separated from the gold. This process drove off any remaining zinc or other waste and left free liquid gold, which was then poured into a mold.

"We have now got the mill running fairly well but most of the machinery being old is always getting out of order. I think within a week or two I will return to the B. Hawk for a month or so especially if Mac comes back. I found it very lonesome there alone and with no help in case of accidents."

A practical man, Del Mar reports that he has his *"chickens up here with me except a pullet I could not catch and the hen that was setting."* Over the weekend he did visit the Blackhawk and found the new chicks were hatching.

Many local items fill this particular letter. Del Mar writes that *"Johnston's black horse was buried with Cactus Flat honors last week."*

And he sends news that *"The Gold Mountain people* [these are the people at mines in Holcomb Valley and the Baldwin Lake area] *will soon have a telephone line to Redlands so we will be able soon to talk with Los Angeles, etc. There will be at least three stages running this summer, Richardson, Kivett, and Knight."*

March 31, 1900
Golden Era Mines

It isn't until this letter that Del Mar explains more exactly where he is and what he is doing. He has left the Black Hawk and is working as a mining engineer for Eberle Brothers between Green Lead and Holcomb Valley. *"We are developing a group of mines and should the prospects warrant it, will erect a mill to treat the ore."*

The missing MacFee has also resurfaced. Del Mar writes that he had a letter from him a week earlier and that MacFee is *"alive and kicking in San Diego County"* and expects to return soon.

About another friend, Del Mar writes, *"Johnston is building a road from the top of Gold Mountain Grade to the Gold Mountain mine."* Comments about the people in the area show how dangerous life was and how diverse were the people. *"Sam Beard had a bad accident. He was a little spirituous and his horse threw him out of the buggy down near Victor. He had a leg broken and his head smashed in. Leach writes me that he is busy with his garden and from his accounts he is planting quite a farm. Mr. Metzgar has gone to San Berdoo* (San Bernardino) *to bring his wife to Gold Mountain and at the same time two young ladies who are doing the country on horseback. We will in time be favored with a visit."*

Del Mar sent a picture with this letter and explains that it shows their log cabin and the mine officials at Golden Era Mines. Pictured, according to the letter, are Del Mar flanked by Charles and Harry Eberle. (The picture has been since lost.)

Remember the chickens that Del Mar left behind at Blackhawk? On a return visit he found the hen that had been setting had hatched two chicks, one of whom died. He brought the hen and her remaining chick to Bear Valley and then to the mine. The weather proved too much for the chick, who *"joined his brother in chicken heaven."* Then he adds, *"The chickens lay five eggs a day and Billy Barnacle* [the rooster] *holds his head higher than ever."*

There is a long gap in the letters at this point. It is likely that these letters have been lost because the letter quoted next tells of Del Mar's receiving her letter dated April 27.

May 16, 1900
Knight's Ranch, Victor, CA

"Mac(Fee) tells me that Leach told him you were coming to Cushenbury this summer. I am glad to hear it for besides the pleasure of seeing you all I long for a piece of lemon pie and a mouth full of cake which I am sure you will give me if I am very good."

He also tells Belle not to forget to bring her song-book entitled *"Student Songs"* and then he underlines the following statement: *"I may be about here this summer and I may not, but I think the former."*

In this letter he gives a glimpse of life in 1900 as he describes the town. *"Gold Mountain is progressing slowly but can hardly be called a town. There are two hotels, two laundries, one barber shop, one butcher shop, and a number of shacks."*

Del Mar has just given a description of the city of Doble, which existed briefly at the base of Gold Mountain in the area known as Baldwin Lake. Baldwin Lake is a dry lake at the east end of Big Bear Lake. Like so many other small mining towns of the time and area, Doble no longer exists except for a small grave-yard.

Del Mar says he expects the mill to be running in two months and adds that a lot of snow has fallen recently so that it is only just then beginning to warm. He ends the letter with hopes of seeing her the following month at Cushenbury.

At this point the letters stop, except for a brief note on June 23, and they do not continue until the following September. There may be lost letters, but it seems likely that for at least part of that time, Del Mar was in close enough proximity to Belle that letters were not needed. The Rogers family was planning its annual trek to Big Bear to camp for the summer. The June 23 letter from Del Mar suggests that prior to the family's journey he must have seen Belle.

June 23, 1900
Black Hawk Mine

This is a brief note that tells of sending Belle copies of the Rubio Canyon photos.

Rubio Canyon is a small canyon that begins on the southern slope of Mt. Lowe and turns to the east and runs behind Echo Mountain where it turns south again and leaves the mountain. The location is much more convenient to Pasadena than to Big Bear.

In 1892 Thaddeus Lowe, an entrepreneur and balloonist, teamed up with an engineer, David Macpherson, to build a railway to the top of Echo Mountain. This line extended the Pasadena tramway system another half mile into the mountains and ended at a large pavilion, which became a major tourist attraction. From the June 23, 1900, note it seems that perhaps Del Mar went to Pasadena before the family left for Big Bear and that during that visit he and Belle visited Rubio Canyon. Because the journey from Pasadena to Big Bear took several days by horse and buggy, it is doubtful that the couple would have gone back to Pasadena once the Rogers family arrived in Big Bear.

The return address on the June 23 note probably explains the lack of letters over the summer months. Del Mar is now at the Blackhawk mine, close enough to Big Bear for the couple to see each other periodically.

September 7, 1900
Victor

The exact date is not known, but sometime between the June 23 letter and the September 7 letter, the young couple apparently became engaged, for now the letters are addressed to *"My Dear Little Sweetheart."*

George Rogers is visiting the mine at this point, and Del Mar notes that he read (part of) her last letter to him. Del Mar states that Rogers is expected to leave for Cushenbury the following morning.

Work at the mine continues, and Del Mar writes that the day after he left her, *"We had the mill going for about an hour to see how the machinery would work. There was sort of an informal reception at the mill, all the natives were there, the Vaders, of course, feeling very big. The tanks had not come nor have they come yet and until I see Vader tomorrow I will not know my future programme* [his spelling]. *He is such a big FIBBER* that I cannot believe a word he says."

He continues with *"I have stayed home from the attractions of the Cactus Club tonight to write to you so you see what a sacrifice I am making for love."* The seriousness of that comment is up to the reader to decide. The Cactus Club, in Cactus Flat, was as close to a local gathering place as existed in the area.

Perhaps to amuse his young sweetheart, Del Mar then describes a trip he had taken recently.

"We went down as far as Mr. Kent's farm, about six miles below Oro Grande. I enjoyed the trip immensely. We just got back this evening. I heard more about horses, cattle and

brands than I can remember and incidentally made the acquaintance of all the hay-seeds along the river. We had some good shooting, 15 of those rabbits like Lotos shot at Cushenbury while your brother and mother were away and which you cooked so nicely, in fact we lived on them and brought three back to Victor for Mrs. Friend. They would not keep to take to Cushenbury. I plainly saw that your father was the friend of all the boys along the river. We had a good time but strange to say we camped without water one evening. The river was dried up and it was too late to hunt for a pool."

September 11, 1900
Victor, California

This is one of the many letters expressing Del Mar's wish to be with his sweetheart. He regrets that her father has now left the mine because having him there was something of a link between himself and her.

On a more practical note, he writes that he is now working regularly at the mill. *"We got the tanks but they came in pieces and have to be riveted together. This will take a month perhaps; it is slow work. I am in hopes we will make a success of the mill. It will be a feather in my cap and will lead to other work."*

Del Mar also tells Belle that he received a telegram a day earlier in which he was offered an immediate and good position Nashville, Arkansas. He did not take it for two reasons: *"First I thought it would be mean to leave Vader at this time and again I did not care particularly to get too far away from someone I love."*

Belle has enrolled at the Normal School in Pasadena where she is preparing to teach school. Del Mar worries that Belle will work too hard, get overly tired, and become ill. His letters to her during her time at the school constantly remind her to take care of her health.

The Normal School was a teacher training college. The idea was born in France and traveled to the United States in the 19th century. The first US Normal School was in the Boston area. A second was in Illinois. California signed on early to the idea and opened the California State Normal School in San Francisco in 1862. It was moved to San Jose in 1870. In 1881 the State Normal School of Los Angeles was established on Vermont Avenue, about twenty miles from the Rogers family home in Pasadena.

In 1919 an act of the California State Legislature made the Normal School the basis of what would become the University of California, Los Angeles. After UCLA moved to Westwood, the Vermont Avenue site was used for the Los Angeles Community College District and the Los Angeles Library system.

Most Normal schools in the United States offered two year courses to train young men and women as teachers. In most states, the student, upon graduation, was granted a Normal Certificate which for many years was a lifetime teaching credential, exempting the holder from the need to pursue additional credit hours to remain certified.

September 16, 1900
Victor, California

Mrs. Rogers, it seems, is the parent most anxious to have Belle, and later Reine, complete the Normal course. Del Mar apparently did not agree.

Del Mar does not wish Belle to become a working woman! *"If I have any success at all I will never allow you to spend 2½ years at the Normal if I can help it."*

"I see by your programme you will have plenty to do, but, dear, do not overwork yourself and lose your health."

Work at the mine continues. *"We are putting the tanks together and will begin to take out bullion in a week or two. The tanks came in pieces and it is a long tiresome job to rivet them together and put them up."*

Reine Rogers has played a joke on Del Mar, which he found amusing enough to relate to Belle. *"She sent me a box of peaches with a note saying she had sent me an jar of jam. I looked all over the box and at last found a tiny little bottle with tomato preserve. Mrs. Friend and I had a good laugh over it."*

September 21, 1900
Victor, California

Belle has evidently written Del Mar regarding an eighteen-year-old young man at the Rogers home. Del Mar assures her he is not jealous and that he trusts Belle will remain true to him.

He reports the tanks are still not ready and that they cannot begin operating the mill until around the first of October.

He has received another job offer via telegram. This one was from North Carolina, and he refuses, as he writes Belle, as he does not wish to go that far from her. Then he adds, *"Should I have a very good position offered me I think you will see me in Pasadena in short order so be prepared at any time for a surprise."*

And he again has words about her studies. *"I*

implore you please not to study so hard as to ruin your health. You probably do not appreciate how many girls ruin their health and dear, for my sake, do not follow their example."

Del Mar writes that he is foregoing the pleasures of the Cactus Club in order to write to her. He also notes that "*the club was trying to get money to buy a piano but could only raise $80 so perhaps they will have to put up with Mr. Martin and Dr. Pitman* [neither of whom is identified further]."

Life was hard and people helped each other in the time and place where Del Mar was located. He writes the following incident: "*A family of a mother and five kids were put off here last Sunday not having sufficient money to go to the Needles where her mother lives. The good people of Victor subscribed about $20.00 and we got her off that night.*"

September 26, 1900
Victor, California

Belle has apparently written a letter asking Del Mar not to take a mining position where he would have to work underground. Here he answers her: "*I will promise I will not take an underground position unless I have to, that is, unless absolutely necessary.*" Indeed, gold mining did not depend on the deep labyrinth of tunnels common to coal and other mining. Further, as an engineer, Del Mar would spend much less time in the tunnels than the regular miners.

Speaking of future offers, Del Mar explains that if he were to take an offer in some of other state, he would want her to come with him. And again, he does not seem very happy with her attending the Normal School.

> *"I wish you would accustom your mother to the idea that you will not finish the Normal 2½ years. At the most you need not go more than a year. If you only knew how I long to be near you, to see you, and be with you all the time you would consider this a long time indeed."*
>
> *"For my sake, dear, do not over study and take care of your health for it is the most precious of gifts. I know you have a healthy soul and hope you look equally as well after the body."*

Life in the town goes on. Del Mar says that he has been eating at the Vader home the past several days but *"will return to the hotel at the first of the month. Mrs. Vader is not much of a cook, and besides not having things I like, as her sample pies, cakes, fruit, and puddings, I do not care much to hear the family quarrels. Besides, Ralph eats like a pig and that I can not stand."*

The mining community was not completely without amusement and culture. In this letter Del Mar tells of a planned minstrel show in which he will take the part of one of the end men.

October 1, 1900
Victor, California

This letter begins on a humorous note as Del Mar acknowledges her correction of his spelling. *"I have just discovered that I have been spelling "sweet" "sweat" in spite of your correction for which I thank you."* And he offers this excuse, *"The word is so new to me that I really believe you were the first person I ever addressed thus."*

Much of the rest of the letter is filled with Del Mar's hopes for Belle and written in rather Victorian prose. *"I am pleased to hear that you are perfect physically…and I know that morally you are perfect so my dear bride to be is as I want her."*

> *"I am glad that my mother wrote your mother. They are both lovely women and I sincerely hope I may never be guilty of any act that will lose their love. Belle, the love of good women is the greatest gift a man can have and I am sure I am well gifted for I have your love and my two mothers."*
>
> *"I should like your mother to get reconciled to the idea of our early marriage for I think I can do well enough to have you with me all the time. My birthday is March 4th and if everything goes well I should like this to be our marriage day, but I am not fixing any precise day."*

October 4, 1900
Victor, California

Belle has written of a nightmare she had in which Del Mar was involved in an accident. He assures her he is fine and that work at the mine continues.

"By next Spring," he writes, *"I will know how I am getting on and we can then consider our future happiness. In the meantime, I will work hard and try to get into a better position. I hardly think that the mine here will keep running continuously but there will be work for at least two months more."* The brevity of a mine's life made it very hard for

a conscientious man like Del Mar to speak in definite terms of a future with the woman he loved.

Until now the letters do nothing to suggest that Belle is not living at home while she attends the Normal school. However, a check of the envelopes of these letters reveal that her address was Cole Grove, Los Angeles. There is no street number. A later one is addressed to 245 Bunker Hill, Los Angeles. In this letter Del Mar speaks for the first time of Belle's being in an apartment, a fact that probably bothered Del Mar.

He writes of a delayed visit to her as follows: *"Mr. Vader went into town so I will have to postpone my visit until a later date. I would dearly love to pay you a visit at your apartment and take you about a little, all by ourselves. Do you cook your own meals or do you board?"*

It was usual for young ladies attending a Normal school to take what today we would call a room but in 1900 was more often called an apartment, in a rooming house. The young ladies were generally restricted to either living at home or living in school approved facilities where only female students lived. Meals might be handled in a variety of ways—the house could provide all meals for a fee (this was called boarding), or the girls might have some kitchen privileges, or each room (apartment) might have a gas plate for limited meal preparation.

Remember Del Mar's telling of the attempt to get a piano at the Cactus Club? In this letter he says, *"We have our piano now. We formed a stock company and paid $150.00 for a square piano. I am billed to sing a song at the Club. I have only seen it once and fear I will break down."*

Surprisingly he adds that he feels *"it fortunate that*

you are away from home now. Your mother will get accus-
tomed to the idea and when we form our partnership she
will not be so cut up."

October 11, 1900 (date not clear)
Groveland, Tuolumne Co., California

Del Mar writes of a long and tiresome journey to
reach his present location. Travel then, like now, was
not always a pleasure.

"It was a tiresome trip up here. There were many
changes and the stop overs were long. From Los
Angeles I went to Merced, from there to Oakdale,
from Oakdale to Chinese and then on stage to
Groveland, a distance of fifteen miles. We had to
cross a bridge over the Tuolumne River. The bridge
had been condemned and the driver warned us. A
married man in the party walked over and as I con-
sidered I was half married I did the same. However,
I think it was perfectly safe."

"The accommodations here and along the route are
rather poor but I like the country about here. The
pines are such a change from the cactus."

Del Mar has gone to this location to meet with a Mr.
Meighan, who, although gone when Del Mar arrived,
was expected back that night. Del Mar is hoping that
he can finish his business that night and get away the
following day.

"I have seen the mine and mill and find it satisfactory," he
writes. *"I am rather anxious to make a good impression and*
get the place. It is a chance I have been looking for."

"The mill is practically in town so I could live at the hotel and have a room as I have at Victor, but perhaps I am building castles in the air."

October 19, 1900
Groveland, California

Del Mar must have visited Belle where she was staying on Cole Grove. He reports here that he arrived back without mishap and he notes that he left her on the car (streetcar) and hopes she arrived home *"without encountering any adventure."*

October 24, 1900
Groveland Tuolumne Co., California

Del Mar has settled into his new surroundings, but this does not include a room at the local hotel. *"I am now somewhat settled although not altogether accustomed to my new surroundings. I have a room in a private house and eat at a boarding house. The hotel people are so inattentive that I could not think of remaining there."*

During his last visit, Del Mar took Belle to a play. Later Del Mar encounters the actors on his return trip to Groveland.

> *"The theatrical people we saw in 'Opium King' were on the same train as I was from Lathrop to Stockton so perhaps it might be interesting to you to know how they look in real life."*
>
> *"All the men look pretty tough. The funny Chinese man 'has a lean and hungry look' with fallen in cheeks.* [The quote is from Shakespeare's *Julius Caesar* and the line refers to the conspirator

Cassius, the man who purposely mislead Brutus into joining the conspiracy to kill Caesar.] *There are two real Chinese, the one who swung the lights and the one sitting on the bench in the last act. The two little girls are his children and speak English perfectly as well as Chinese."*

"The Opium King seemed to be very sweet on Georgette who is a rather fine looking blonde but with not much style. The old Irish woman is the same coarse creature in real life. The Opium King's wife is a woman past her prime, rather too lively and has a cracked voice. The bicycle girl is a young girl about 20 rather pretty with a fine head of blonde hair, rather lively, and, I should think, independent. She wore a long light coat and street sweeping skirts."

Del Mar is preparing for his new position. *"I have begun work mapping the mine and then will assay all the workings before taking actual charge of the mine."*

October 30, 1900
Groveland, California

A sentimental side of Del Mar comes through in this particular letter. *"I thought you did well when we parted. I knew how badly you felt for I felt the same. It is not that I don't like crying people but that I do not like to see them crying."*

After more words of endearments, Del Mar says he is *"looking forward to the Spring and the happiness it will bring us, so don't study too hard but just enough to show you are not the least brilliant in your class."*

As for his work, Del Mar notes he has mapped the mine and is getting ready to assay the samples. *"It is a big property and there is a future for me by making it a success under my management. The present manager does not leave for at least a month and I think I can do better than he is doing."*

November 4, 1900
Groveland, California

Del Mar here responds to a letter Belle has written regarding what she must have termed "the practical side of teaching." Del Mar, still not anxious to have her pursue this career, responds, *"The more you hear about the practical side of teaching the more you will appreciate the fact that it is hard work,"* and then he adds, *"but I hope it will not be too hard teaching me to always be your lover."*

Always one to observe the rules and decorum of his time, Del Mar now tells Belle, *"I know I have done wrong in not speaking to your father but he never brought the subject up and I always felt too embarrassed to mention it myself. I will write him explaining the difficulty."*

And what must have been good news for Del Mar, he answers *"I like your idea of leaving Normal at Xmas if your mother is agreeable. I promised your mother I would be content to wait for two years and this promise we must keep unless she is willing that it should be broken, but between you and I and the gatepost I never thought we would have to wait that long. If your mother will give her consent willingly, by all means leave after Xmas and take a little rest, then in the Spring we can build our little nest together and next Summer we can have the family up here to visit us."*

It is not clear why it is the mother's permission, not the father's, that Belle seems to need before she leaves school. This might be in part attributed to the fact that her father was gone so much, leaving the mother in charge of the family.

Now Del Mar tells us a bit more about the area where he now works. *"I have not mixed up with the people here but as far as I can see the place is made up of old time bachelors and young couples. In fact, it is a young camp in more ways than one and although occasionally the boys get a little jolly and ride into the saloons, yet on the whole the place is rather quite orderly. One seldom hears of an arrest; in fact, none since I have been here."*

"The town of Groveland," he continues, *"is situated in a valley with hills on both sides. The grass even now is green. The climate is pleasant and the Winter moderate with but little snow."*

Work, he tells her, has been interrupted by a broken pump. The mill had to be closed while repairs were made.

The first mention of anything political is in this letter. Del Mar writes, *"We had a Democratic meeting and some speakers and it was most orderly, no drunks, no interruptions, no shouting, especially as it was a Democratic meeting in a mining town."* Such meetings in mining town must have had a reputation of being less than orderly.

November 8, 1900
Groveland, California

Belle has written of problems with her eyes. Del Mar takes this chance to encourage her to stay at home. *"I am exceedingly sorry to hear your eyes have broken down so soon and hope a little rest will be sufficient to right things."*

> *"If you love me, Belle, do not return to Normal and ruin your health as it surely will. Do not live to regret the acquirement of a little knowledge which you can acquire at leisure. Stay at home till next Spring and learn from your dear mother all the little duties and arts that tend so much to bring happiness into a home, study your music and read good healthy books."*

The letter goes on to describe how he will come for her next spring and they will have their own home. *"There is always plenty to do in this world and let our lives be pure and good."*

Now Belle was in an awkward situation. She is engaged, but her fiancé is not present. She is a young woman, and staying home all the time is not attractive. Custom and manners of 1900 put some restrictions on her. However, she has cousins and friends who would like to have her join in activities with them. She has mentioned this to Del Mar, who responds, *"I do not in the least mind if you go to the theater or elsewhere with your cousins or anyone else and I appreciate your telling me the fact."*

> *"Your troubles will come as everyone has to experience a certain amount and now while you are young, get all the healthy enjoyment you can. I regret very much that I am not stationed near you to take you about as I know an engaged girl is rather at a disadvantage."*

A new method of communication is mentioned in this letter—the telephone! Del Mar says that if he can not be with her at Christmas that perhaps they can arrange a long distance call. No mention is made of where he will access a telephone or if there is one in the Rogers's home.

Now Del Mar turns back to life in Groveland. *"We had a heavy rain yesterday and gum boots were the order of the day."*

"The manager here is a little difficult to get on with, but I am getting on to his 'curves' and know better how to take him. He is rather prone to forget his orders and then give different ones, but so far we have had no friction."

And finally, he speaks of the recent presidential election: *"It is too bad we lost the election but beyond feeling that I am in the minority in my political opinions, I am not hurt in any way and know I can live under McKinley as well as Bryan."*

November 8, 1900
Groveland, California

In the previous letter to Belle, Del Mar commented he had received a letter from her mother on the same day. What follows is from his response to Mrs. Rogers' letter.

My Dear Mrs. Rogers,
"Your letter informing me about Belle's eyes was received. I was exceedingly sorry to hear she was at all sick and hope a little rest will be sufficient to right things. I think Mr. Rogers was right in wanting

Belle to come home, at the same time I know you have her welfare at heart and would not let her return if you thought it would hurt her in any way."

"In my opinion and for me, Belle has had a good education that is a good groundwork to build upon, for this is all a school education can give and now by a careful selection of good books, she can build that foundation to any height she has a mind to. If her particular liking is music or any other art or a science, I think providing you can afford to let her do so, it will best to allow her to pursue it at leisure. If I had the choosing of a wife over again, I would still ask Belle to join her lot with mine and I hope next Spring you will give your consent to our marriage."

Later in the letter he says, *"I consider Belle's health of far more importance than getting a teacher's certificate. At the same time I fully understand the advantages connected thereto."*

"I appreciate that it is a disappointment to you that Belle may not be able to finish at the Normal, but at the same time I want you to know that as long as I live and have health Belle shall not want the refinements and necessities of life and that I fully expect within four or five years to give her a home as prosperous and happy as any in the land."

Del Mar closes the letter with this request to Mrs. Rogers: *"While Belle is at home both she and I will appreciate your teaching her all the little duties that help to make a happy home."*

November 17, 1900
Groveland, California

Del Mar begins this letter by again expressing his approval of Belle's decision to not return to school. For his part, he writes of wishing she were there and also saying that he would not ask her to settle in Groveland in the winter.

"We have had two days of steady rain with no signs of stopping. One has to go about in gum boots. It may interest you to know that we have a butcher shop, two groceries and a few saloons, also a hotel and a restaurant. While the manager is here I am under his orders. He expects soon to go East for which I will not be sorry. He is a well meaning man but likes to show his authority."

In this letter Del Mar mentions he has not had time to write to Miss Stevens and will do so soon. There is no explanation as to who this lady is.

November 23, 1900
Groveland, California

After professing his love and concern for her, Del Mar turns to more mundane items. *"We have had four days of rain. The roads are very muddy and the stage was delayed one day, perhaps the day I wrote to you."*

That last was probably in response to Belle's last letter in which she apparently had complained she had not heard from him. She had apparently also asked if he were ill and was that why she had heard nothing. In regard to his health, Del Mar writes *"Cheer up, my dear, and don't be sad or weary or I will have to journey south to see you and lose my position in so doing."*

In the rest of the letter he contemplates their future life together: *"The time is passing quickly, however, and it won't be long before we will have some sort of home of our own, not a palace perhaps, but a cottage full of love and sunshine."*

November 27, 1900
Groveland, California

In this letter Del Mar tells us more of his work at the mine. *"I have been working pretty hard lately putting up a new pumping plant to force water up to the mine from the creek."* Getting water to the mines is a constant concern for the miners in the mountains.

The weather has cleared after the storms. *"We can see snow on the mountains two miles away, but have had none here so far."*

"The fare at the hotel," he writes, *"is very poor indeed and I am looking forward with great pleasure to the time when we will go to housekeeping and can have what we want to eat."*

"Thanksgiving is near at hand," he notes, *"and the poor turkeys are having a hard time and some poor boys will likewise have nightmares and colly wobbles."*

Please note the last few letters and those that follow for several weeks are definitely of a man very much in love, almost desperate to see his sweetheart and make her his wife.

"I will give thanks to the Lord for having found a dear, good, sincere, and congenial sweetheart and I

hope that she will shortly be my sweetheart, friend,
and wife and will always remain so. If I could find
a suitable house here I would be tempted to ask you
to come here sooner than next Spring for I love you,
dear, and want your sympathy and affection."

November 29, 1900
Groveland, California

Much of this letter is devoted to the fact that each
is complaining about letters not arriving. Del Mar says
he has written three times since he last heard from her.
He asks her to write as soon as she receives this letter.
"I love you so much, Belle, that I would leave no stone
unturned to get word to you. Should my letters not reach
you, you would be justified in concluding that my love had
grown weak but no such thing has happened for I love you,
dear, as I love no one else."

One negative note, almost a warning, comes near
the end of this letter.

"I wish I could spend Xmas with you but as the
Manager goes to Frisco on the 17th I will have to remain to
'keep house.'"

December 2, 1900
Groveland, California

The mail has finally gotten back on track for the two
lovers, and letters are arriving on schedule. Now Del
Mar reassures his sweetheart of their future when he
writes, *"You must not imagine that you are going to feel*
homesick with me for you will have a home of your own with
all the love and sunshine that is in my power to give you."

And he continues by telling of his plans for the future. *"My present intentions are to stay here until I find something better. It will take at least five years works at mines before I can hope to set up a business in the city and as I enjoy the mountains so much I may remain longer, that is if you are of the same mind, for when we enter into our partnership the final decision will rest with you should we not be of the same mind."*

MacFee is mentioned briefly here. Belle may have written something about cooking apples for Del Mar answers, *"The apple cooking is just like Mac. He believes too much of what he reads. I have only heard from him once and wrote to him yesterday."*

The telephone call that was mentioned earlier is now discussed again. *"I would dearly love to see you at Xmas but it is not possible. Find out where in Pasadena you can speak through the long distance phone and at a certain time Xmas morning to be agreed upon you be at the phone and I will try to get a connection with Pasadena. We can then have a couple of minutes talk."*

Again Del Mar talks of Belle's coming to Groveland. He has decided, though, that he won't ask her to come before spring (about March 3rd). *"However much I should like you to be with me sooner for I know you must have time to get accustomed to the idea, but I think you will never regret your married life. My first duty will always be your happiness. My second will be to be happy for your sake and my last duty* [will be] *to make your love grow stronger every day."*

Here Del Mar admits he still has not written George Rogers and asks if he is on the desert at this time.

For quite some time Del Mar has not mentioned his

camera. Now he says, *"My camera is having a rest these days—no time to give to it. I am all the time looking out for a good position in the Southern part of the state so as to be near you but have had no success yet."*

December 5, 1900
Groveland, California

Throughout this time Del Mar has kept a fairly steady correspondence with his mother and other family members back in New York. Now, however, he complains he has not heard from any of them, not even his mother, in a month. *"Surely,"* he writes, *"they can not think because I am engaged to you that I do not want to hear from them any longer."*

Apparently there is some social life in Groveland, for Del Mar writes, *"They had a dance here a few nights ago and kept it up till 6 A.M. I did not attend but saw the couples leaving as I went to breakfast. There will be a mask ball at Oak Flat (a mile and a half away) on Xmas eve. I will not go but you can imagine a mining town in mask or perhaps the imagination can not go so far. On ordinary occasions the boys go out to liquor up between each dance and by midnight the room smells like a whiskey keg. I presume on this occasion they will come already well charged."*

The social life offered in the town apparently has no appeal for Del Mar. Whiskey and other spirits were not to his liking, an unusual case for his time and place.

He continues, *"I have made very few acquaintances about here and have no special desire to do so. I am a bit lonesome at times and want only you to cheer me up. At these times I long to see you and have you near me and then I think, well, I must not be so selfish as to want you to come here during the winter, which, after all, has not been a winter—a little rain and not very cold."*

Perhaps to tell her more about the town, he *adds,* *"There are lots of Italians about here, in fact all the hay-seeds are Macaronis."*

December 14, 1900
Groveland, California

Here he thanks her for the violets she sent and says they still retain some of their scent. What follows is a discussion of his name and how he would prefer she address him. Mentioning that his own names (Algernon Percival) are not familiar ones to most people, he says that in the past some friends called him Sammy. Then he says he prefers *Alg* to *Algie* but that she may decide herself what to call him. He always signs his letters as "Alg".

Now for an update on what is happening at the mine. He writes, *"I have been very busy this week, the Manager being away in San Francisco. I told you some time ago that the Manager was difficult to get on with. I think it my duty to tell you that although we have had no words yet, I am not wholly satisfied and that I am looking for a better place. As soon as I get a firm offer I will have it out with the Manager so that if things are not changed a bit, I will have another place to go to. I have now an inquiry from Oregon. As you say, a steady position is best but one must be satisfied."*

This must have been difficult for Del Mar. He wanted a position that would afford the stability he needed to marry Belle, and yet, on the other hand, he once again finds himself in a situation he feels is wrong for him.

Del Mar repeats his wish for Belle to be ready by March 3rd but will defer to Mrs. Rogers regarding this date. Remember he said earlier he would like to have his March 4 birthday be their wedding day.

A few details about Del Mar's life at Groveland emerge in this letter. He spends each evening until 8 p.m. reading and writing letters. Then he goes to bed in order to rise early.

And that Christmas telephone call is once again mentioned. *"I think the best time to get the phone will be Dec. 24. Let's make it at noon on Dec. 24. You be at the Pasadena end and I will try to call you up. Let me know your number by next mail."*

December 20, 1900
Groveland, Tuolumne Co., California

The following letter throws all the plans up in the air.

"A surprise is in store for you. I have had it out with the manager and in consequence I will leave the company. This may be bad news to you and, if so, I am sorry for your sake. I have not been given my walking papers but we have come to an understanding that we do not suit each other. He is a most difficult man to get on with and has not been able to keep anyone in his employ. We may as well have our talk over the phone as I will be here but will journey south soon after so that I will see you on New Year's Day if nothing happens. I intend to see a little of the country about here, that is, the mines on the mother lode."

"Of course, it is a disappointment to me that after I had thought to be a little settled here to have to pack

up and move, but, my dear, I have such faith that you are my mascot that I have no fear but that better things are in store for me."

"The lesson I have learned is an old one, i.e. never to work without a contract in writing. The whole trouble is that I came here to do a superintendent's work and I was wanted to do a mechanic's work."

"It may be possible that I may stay here for a month or so but I do not anticipate so doing at present."

"Now is the time, my love, that I want your consolation for in spite of the fact that I am leaving of my own free will, yet I am disappointed that I will have to change my plans for next summer. I had anticipated a trip to the Yosemite but that can not be now."

Interestingly, there is no mention of any possible change in marriage plans, and the letter ends with talk of love for her.

December 27, 1900
City Hotel, Sonora, California

Now in Sonora, not far from Groveland, Del Mar writes of the problems that led to his leaving Groveland.

"I was too badly treated to be able to stand it longer. You read the letter I received from the Manager when I was in Pasadena. I will only give you two examples of my treatment."

"The 21st of December I got up at 2 A.M., ground a valve on a boiler and walked five miles in rain to get a fitting. This all before breakfast. Then I worked all day on a pump until 1:15 the next morning or in all

53

nearly 24 hours. On the 24th Dec. I worked running a pump from 5 A.M. until 10:30 at night and then, because I asked for a man to relieve me, I was talked to by the Manager as if it was my duty to keep on. As a matter of fact, I did keep on until 7 A.M. the next morning (Xmas) just to spite him and then I asked for my time."

"When I quit the amalgamator did the same as well as the concentrator man so that in spite of his trying to keep the mine going on Xmas day everything was shut down."

Reading these two incidents makes one wonder what else the rather tyrannical sounding manager may have done to the men. It is no wonder Del Mar was anxious to leave Groveland.

"If nothing comes my way by the 2nd of January," he writes, *"you will see me about the 5th. I will telegraph you on the way down when if you will kindly find out where I can get a room in Pasadena as I want to be with you for a week or two."*

And with this letter, the first year of the engagement of Belle Rogers and Algernon Del Mar comes to an end. A year that began with high hopes and a romance expected to end in marriage finishes on a low note as the means of support Del Mar hoped would enable him to marry his beloved Belle has abruptly ended.

1901

According to his work diary, in 1901 Del Mar became acquainted with Mr. Johnson (no first name given). Johnson had recently sold a turquoise claim in the Clark Mountains to Tiffany Brothers, the New York jewelers. The sale left Johnson with ten thousand dollars that he wanted to use to find a mine. Del Mar agreed to help him in return for three dollars a day plus expenses. Further, if a mine were located, Del Mar was to have a half interest in it.

Almost immediately there were problems, which are mentioned in the letters that follow. Johnson took in another partner, a promoter named Robinson, without consulting Del Mar. It also became apparent that Johnson was a heavy drinker, a trait of which Del Mar was critical in anyone. Fortunately, perhaps, for Del Mar, the new partner soon dropped out, apparently in part due to Johnson's drinking.

The following letters tell us Del Mar's actions as well as his thoughts about his arrangement with Johnson.

February 27, 1901
Manvel, Nevada

Del Mar first sends Belle a penny postcard reporting he has arrived safely at Blake. A letter written the same day as the postcard tells us Del Mar was traveling from Manvel to Sandy. Jobs were not easy to find, and good jobs that paid for Del Mar's training and expertise were even scarcer. The result was a lot of traveling, much of it done on foot, horseback, or by wagon. Most mining areas were not directly accessed by rail or good roads. The Feb. 27 letter from Manvel describes the settlement of Manvel (in Las Vegas desert area) as *"a sandy waste clotted with canvas habitations, not very inviting to the eye. I think death would be preferable to being situated here."*

This letter is one that also gives some specific details about travel and life in that time. Having one's baggage fail to arrive at the right place and the right time is not a modern inconvenience. Del Mar reports that his assaying material and tent have not yet arrived and no one is sure when they will get to Manvel. Traveling from Marvel to Sandy is not inexpensive for those who must hire transportation. *"The stage charges $8.00 to go to Sandy and teams cost from $15 to $20 for the trip. There seems to be quite a little business here in teams, at least on train days,"* he says.

Remember, please, that the trains of 1901 were driven by steam. A train crew needed fuel (wood or coal) to heat the water to make the steam, which in turn drove the engine. Sometimes things did not work perfectly.

In describing the train trip to Manvel, Del Mar writes, *"The train from Blake went through its usual per-*

formances. When we got about 4 miles from Blake the water gave out so the engine had to run ahead four or five miles to get water and returned to the cars, then the steam gave out and we stopped for half an hour to get up steam and go on."

March 6, 1901
Sandy, Nevada

Del Mar is now in Sandy Valley of Clark County, Nevada, about forty-five miles from Las Vegas. Sandy, now called Sandy Valley, is in the Mesquite Valley. It is bordered on the east by the southern extension of the Spring Mountains and on the west by the California state line. There were actually five mining towns that were started in this area in the 1880s: Kingston, Ripley, Sandy, Mesquite, and Platina.

Del Mar reports just having been to see Robinson's *"Big Iron Dyke about which Robinson talks by the hour…and even when we had camped in sight of it he informed us that we would be surprised in the morning and so we were for after going over the property Johnson and I decided there was nothing worth putting in a cent and so we left Kingston Mt. All the talk of excitement at Kingston is rot."*

Del Mar and Johnson plan to take a trip to the Colorado River while Robinson is away. Alg says Robinson is calling people aside every time he meets someone, so now Del Mar and Johnson have privately nicknamed him *"Iron-Dyke-call-him-aside-Robinson."*

He and Johnson, Del Mar admits, are getting pretty tired of Robinson and plan to make arrangements for him to leave so they can do the mining and Robinson will be responsible for only the financing, since his enthusiasm runs away with his judgment, and they feel they can't believe a word he says about a mining prospects.

March 8, 1901
Sandy, Nevada

Del Mar reports that the name *Sandy* is well applied where he is. *"The town is made up of a store and saloon, four or five tent houses and the old ten stamp mill is near the center of a large sand sink. The principle lady and the only one here is the Indiana wife of the proprietor of the station."*

They plan to go to the Kingston Mountain the next day and then decide whether to make their headquarters there or in the town. *"There is a little boom here. Some fortunate men are at meals everyday. Prospectors are asking thousands for mere prospects."*

"The disintegration of the Robinson-Johnson Co. has already begun," Del Mar writes. *"R., no doubt, is a great talker that J. is tired of him already. I see a break up not very far ahead. I sometimes admire the way in which he can talk hours on something that could be done in minutes—that is I admire it when he is not practicing on me. He certainly is tiresome, as I told you some time ago."*

March 8, 1901
Sandy, Nevada
(these last two letters have the same date)

Del Mar tells Belle he is writing every time he can so she won't miss him so much! In this letter he tells of the personality conflicts with Robinson and his own plans.

"Robinson left yesterday for Manvel but had to give his place to a man at Goode Spring some 14 miles from here. He must have felt very cheap. It is a great relief to get rid of him. The Robinson, Johnson, Del Mar outfit is just about at an end. Johnson and I will take a trip to Vegas and about the Colorado River." He expects this will take ten days to two weeks, and on their return they expect to assay their samples. If the assay results are satisfactory, they will go back in business. If they see nothing, they will go to the Bluebird and then to Randsburg.

Del Mar has pretty well tired of Robinson and writes that he intends to have nothing more to do with the Robinson. *"Mr. R.'s reputation is not of the best and his companionship does me more harm than good. Everyone here makes fun of him."*

"He [Robinson] *heard here that the Copper World had shut down and that the mine had been sold. On the last information he got very sick and left here looking a wreck."*

Del Mar's letter contains a puzzling statement: *"I had never given a thought to the leave taking at Pasadena and as far as I am concerned I left as if I would return tomorrow for the same reason that your mother did not come to the door."* One might suspect that his leave-taking from the Pasadena visit was without the fanfare that Belle might have thought appropriate.

Then the letter goes back to business.

"I find here as usual, after driving 15 or 20 miles and walking to the top of a mountain to see a prospect that had been highly praised to find it not worth putting a notice on. I am not favorably impressed with the mining prospects here and will be

glad to get to something more substantial. I am not discouraged but I expected nothing more. Call-him-aside as I told you in Pasadena is a regular gas bag—all talk."

March 11, 1901
Las Vegas, Nevada

Del Mar, Robinson, and Johnson have gone into Nevada searching for likely spots to prospect.

"Our programme now is to look at some prospects near here then return to Sandy via Manvel or White's Ranch as it is called. Robinson has gone in and we have got rid of him I hope for good. He is a crank, an enthusiast, and has no judgment. After looking at various prospects he picked out two or three on which he thought he could get in his work. He explained it as a fissure vein cutting across the limestone formation. I went to see it and found it a chloride prospect and one on which I could not make a favorable report. There is not a mine or prospect that I have seen that I would take the trouble to put a location notice on."

"From Sandy Johnson and I will take the assay outfit to Manvel and ship it to Randsburg. We will go by team to the Bluebird Mine and if we think well enough of it, we will make some arrangement with your father to get a bond on it for six months. If your father is at home tell him I think he had best give me the power to make an agreement something in this line."

Now Del Mar sets the outline of the way he feels the agreement should read. *"First we have six months to do work on the claim, that no ore be shipped and that at the end of six months if we want this claim we will pay $2000 and six months later $3000 more."*

The travels continue. *"From the Bluebird we will go to "The Caves" and from there to Randsburg where Johnson knows of something that was reported well to him. On this trip outside of the Sandy and Good Spring District, Robinson and I part and Johnson and I go in on equal terms, he putting up the dough."*

"If we find nothing to suit us at Randsburg I will either return to Los Angeles and go to Virginia Dale with Robinson or return with Johnson to Sandy and take a trip into Nevada with J. and another man who knows the country."

In the coming travels, Del Mar hopes to meet Rogers on the road to Manvel and Lavic. And he lets us know that he is still not a big fan of "Call him aside" Robinson but that he might be helpful in the business and the three of them might get a hit.

The description he gives of Las Vegas is a long way from the glitz and glitter of that city today. *"Las Vegas is a ranch on the east side of mountains in a large valley (vegas). A spring about the size of Bear Creek (last year) covers the ground a little above the ranch."*

March 13, 1901
Manse, Nevada

At this point Del Mar is traveling daily. as the following illustrates.

"You will receive this letter at the same time as the one written from Las Vegas. From Vegas we drove 22

miles to the Cottonwoods and today some 50 miles from the latter place to here (White's Ranch). Tomorrow we go to Parump and return at night. Saturday back to Sandy. Monday we go to Manvel. Tuesday to Fenner, Wednesday to Lavic, Thursday to the Bluebird, Friday to Lavic and Saturday start for Randsburg."

Because he is on the move here, he tells Belle he is having his mail sent to Pasadena once again and asks her to send it to Randsburg until March 16th.

As for his present location he writes that *"Manse or White's is quite a farm out in the desert. About 100 acres are under cultivation with alfalfa, pears, and fruit of all sorts."*

March 17, 1901
Manvel, Nevada

Del Mar writes Belle that he has written Robinson that their arrangements is no longer in place. He is currently back in Manvel, he tells her, and traveling to Randsburg to the Bluebird and other mines. He reports receiving a letter from George Rogers (Belle's father). Del Mar apparently sent Belle a new parasol and writes that he hopes to help her use in a few weeks. He also gives her *"my full consent to go to Mt. Lowe."*

Mt. Lowe was a resort on the foothills of Alta Dena that was reached by taking the Big Red Cars of Pasadena to Alta Dena where there was an incline cog railroad that took passengers to an observation point which was Mt. Lowe. It was the place to go at the time. Del Mar says he has full confidence in her and knows she would not go with anyone to whom he objects.

March 21, 1901
Ludlow, Nevada

The following letter tells of an all too common problem with travel through a desert. Del Mar and Johnson have survived a very unpleasant experience, which he describes as follows:

"We left Manvel at 8:50 a.m., rested at noon at Restler's Springs (24 miles) then on to Marl Spring about 10 miles. We brought a 10 gallon keg with us to take water from Marl Spring to Mesquite Springs some 30 miles."

"Tuesday the 19th we left Marl Spring early, just before noon we found the bung had come out of the barrel and that we had no water. We were all right if we struck the spring that night. We hunted until dark but could not find any. The horses were without water all day and we since noon. We started out next morning early but had no better luck so there was only one thing to do and that was to unhitch the horses and make for the railroad. We made a 20 mile walk without water but took a few tins of string beans from which we got some moisture. We struck the railroad about 2 miles west of Klondyke to where we walked and put up for the night. We left our wagon and everything on the desert and will now have to strike out from here to find it. We walked from Klondyke to here today 12 miles as we heard there was an old wagon road going near where our wagon is. It was a close shave, my dear. I thought of you and determined I would not give up. We did not ride the horses as they were without water for over

30 hours and were about played out. Tomorrow we will prospect from here for a horse to bring the wagon through."

"Klondyke is a box car occupied by two telegraphers who treated us very nicely but here at Ludlow they are very disobliging. We asked for some lunch about 12:30 but were refused even a piece of bread and butter. We have nothing with us except our clothes, the two horses, and some money."

March 27, 1901
Randsburg, California

Del Mar and Johnson are now approximately forty-five miles from the Las Vegas area, working out from Sandy via Manse or White's Ranch.

Robinson has finally left, after displaying what Del Mar describes as his crankiness and lack of judgment. At one point Robinson picked out two or three spots he thought might work. He explained one location as *"a fissure vein cutting across the limestone formation."*

Del Mar examined the spot and found it was a fluoride prospect. He writes, *"There is not a mine or prospect that I have seen [here] that I would take the trouble to put a location notice on."* Remember, a location notice is the preliminary step to filing a claim.

Del Mar is continuing to travel with Johnson and writes that he will take the assaying outfit to Manvel and ship it to Randsburg. He speaks of going with Johnson to the Bluebird Mine. He concludes that if Johnson thinks Bluebird okay then they will contact Mr. Rogers to get a six month bond on the place. The six month bond is a short-term way to hold a claim, and at the end of that time, more serious money is due. Del Mar repeats his earlier intention that at end of six months they would pay $2,000 for the claim and then an additional $3,000 six months later.

Del Mar recounts the travels he and Johnson (and Robinson before he departed) did to look at places (mines) in Randsburg, Sandy, and the Good Spring District. His path was likely to cross that of George Rogers, and Del Mar often mentioned the possibility of encountering Belle's father on these travels on the road from Sandy to Lavic. All the towns where Del Mar is then working and traveling are near Las Vegas. He describes Las Vegas as *"a large ranch on the east side of the Clarkston Mountains -a spring of water about the size of Big Bear Creek last year comes out of the ground a little above the ranch."*

March 28, 1901
Randsburg, California

By now Belle has returned from her trip to Mt. Lowe and writes that she has a cold. Del Mar consoles her about her health and then noted that is it very cold and windy where he is.

"We came from Ludlow via Lavic, Peacock Wells, Daggett, Barstow, and Johannesburg. Although we were at Peacock Wells we did not go to the "Bluebird" as Johnson was in a hurry to get to Randsburg. As it turns out we were too late. The option on the G. B. Mine gave out the night before so that we were too late. This is the property we were to examine and if favorable to put a mill on. Johnson says he will put up a few hundred dollars to do the

work on the "Bluebird" on my recommendation. When we get to town we will see your father and make arrangements to develop some. The Peacock will soon start up again. They will concentrate the ore and ship the concentrates. The fact of the Peacock starting again will add value to your father's prospect."

"The mines that we have seen so far about here have been gouged out so that they are in very poor condition to work on. We may be in Los Angeles in a week unless we find something here that looks well enough to work on."

Remember that Del Mar is also an amateur photographer and as such often asks her to send him his mail and lists some photographic supplies he wants sent. This list includes a dozen films for 5x4 *Kodac* (his spelling) (not 4x5) which he says she can get in Pasadena for ninety cents.

"Call-him-aside [Robinson] *has gone to Arizona I think. At all events I have broken all connection with him. He is both a knave and a fool as well as a crank."*

March 31, 1901
Randsburg, California

Three days later Del Mar is headed toward Los Angeles from Randsburg. He writes of snowy weather and his plans to travel from Randsburg to Mojave via Garlock and then to Rosamond, Lancaster, San Francisco, and finally to Pasadena. He tells her that reports say there *"will be water all the way and no fear of being lost."*

April 12,1901
San Luis Rey, California

In a short note from San Luis Rey Del Mar reports that they had taken many wrong turns and are now on the way to Escondido. He is apparently photographing places on the road as he asks Belle to send a roll of films (5x4) to Temecula and have it kept there ten days.

April 21, 1901
Temecula, California

The men have reached Temecula, traveling by horseback. They have traveled thirty miles since beginning at 6 a.m. and have stopped to rest their horses a few hours. *"We got lost again—in Deer Park trying to make a short cut from the mines to Cuayanaca, but we turned on our tracks and lost half a day thereby. We were in a well watered country so did not mind."*

June 5, 1901
San Bernardino, California

There is no mention of why the big space between letters—perhaps some were lost or, more likely, he was located close enough to Belle that they were able to spend time together.

Del Mar has been exposed to the mumps but apparently has not contracted them. He also reports Chief (his horse) has colic and that he plans to rest him before going on.

Remember, an earlier letter says that Del Mar plans to ask Rogers to go to the recorder's office on his behalf. Now Del reports going *"to the Recorder's office but could find no trace of Bluebird Mine"*! Then he continues, *"If your father comes in time for Thursday night mail ask him where he had the record made of the claim and send us the copy from the Recorder to Daggett."* And then he adds, *"We must know about the record as soon as possible."*

Now Del Mar wants future mail to go to Lavic via Bagdad.

June 8, 1901
Victor, California

In this letter Del Mar tells Belle he is back in Victor and that stories of Vader are prolific still. *"Everything* [with Vader] *is oil. I have had lots of fun joshing the boys about here over their oil companies."*

Most of the letter concerns horses. Del Mar says he has written Rogers and recommended that Chief (the horse who had been sick) be pastured and then brought to the mountains later. He recommends not selling the horse despite Johnson's having told him (Rogers) to do what he wished with Chief. *"J. was a little spiritual at the time,"* he explains.

Del Mar and Johnson are going at a slower pace because the horse that replaced Chief is slow and old *"something like our old white horse at B*[lack] *H*[awk] *(whitewings)."*

June 11, 1901
Lavic (site of Bluebird Mine), California

The men are now back at the Blue Bird in Lavic. They are ready to take their equipment to the mine since all supplies ordered for the mine have arrived.

Del Mar is now referring to the horse as *"an old*

unadulterated plug." He describes high winds that are cooling the atmosphere a bit.

"The people at the well," he reports are *"affording us every facility for work in the way of water, etc."*

Included here are some details of his travels that indicate the time consumed by travel in 1901. *"The first day out from San Burdoo we camped in the Cajon. Then Saturday to Victor where we stayed all day. Then to Daggett. Then Monday to near the Peacock Well. Then today to Lavic."* The total distance covered is around seventy-five miles.

June 13, 1901
Lavic, via Bagdad, California

"We are now established in Bluebird Canyon and will begin work tomorrow on the mine."

One interesting part of this particular letter is the updating of the people who live and work in these early mine and the towns around them. *"Bruce Johnston is working at the Rose* [in the Cactus Flats area] *and according to the old man he is a fine fellow and is saving money fast. Mac* [MacFee] *is out on the Desert but no one knows where. Preciado has the mail contract to Gold Mountain but Richardson will still run the stage."*

He asks Belle to tell her father *"that Eliott has got his walking papers from Shrader who is now superintending."*

Water is a necessity, and arrangements to have it at the mining camps and mills needs to be made since in most instances these mines are in desert areas. Del Mar relates here that *"The Peacock Co. will charge us 50 cents a barrel for water but nothing for watering at the trough."*

Next Del Mar describes finding a stray horse, which he describes in some detail, including a brand, as he thinks it may belong to Rogers.

"Joe [?] saw us stop at Garri's but you were right, I did not go inside. One can stop there without suspicion as it is the only watering trough for some distance."

He estimates their location as 2,000 feet above Daggett.

No Date
Lavic, via Bagdad, California

This letter appears to follow the one above. Del Mar writes that camp is now established and describes it as two tents with a large piece of canvas stretched between them to give them shade all day. *"The wind comes up about 10 a.m. and from then on it is cool in the shade but hot in the sun."*

Two other men are at the camp while Del Mar is writing: Johnson, who is playing the violin, and Fred Sears, who is lying on the ground.

There has been a new addition to the group—a dog. *"We have a fine dog we got at Victor. It is a cross between a pointer and a setter. He was called Mace but at Stoddard's Well we baptized him in the tank and called him Nig."*

As for work, Del Mar says they have been quite busy and this has kept him from thinking all the time of her, but now that things have settled down a bit he realizes how much she means to him and how he loves her. He asks Belle to report to her father that the work so far *"has been encouraging and they intend to run a tunnel right to the hill from the lower workings on the side hill."*

June 21, 1901
Lavic, via Bagdad, California

Belle has the mumps. Del Mar writes that he wishes he were there to console her and mentions that since he has had no symptoms so far he feels he will not get the disease.

Because he moves around so much, Del Mar frequently has to use storage facilities to store his belongings, and he depends on Belle to take care of a number of items relating to his mail. In this letter he asks if she has put his trunks in storage and instructs her to have all papers at the post office sent to him in Lavic. The Rogers family will be leaving soon for Big Bear, and here Del Mar reminds Belle to make sure the Pasadena post office has his Lavic address. He sends a dollar to pay postage plus purchase two items for him. He wants *The Student Song Book* and an autoharp instruction book. He also asks her to send any extra newspapers or magazines she might have. He is especially looking for a magazine called *Argosy*.

Del Mar confirms that Fred Sears is with them and that he met them in Victor. And the work goes on: *"It has been very hot these last two days. We are running a cross cut tunnel to strike the ledge and it will take about two months to get in that far (about 140 feet). Mr. J. is not at all discouraged that the surface indications did not extend far. I think we still have a good chance."*

June 30, 1901
Lavic, via Bagdad, California

Del Mar immediately tells Belle that he forgot to tell

her that he had enclosed in his previous letter a deed. He now has one-third interest in the Bluebird mine and wants her to keep the deed which shows his ownership.

Once again the horse named Chief enters the picture. *"If your father takes "Chief" to Cushenbury or Bear Valley as I before planned, I want the exchange of "Chief" for the grey horse we now have as an excuse for me to go to Bear Valley. When you get to Cushenbury and are about to leave for Bear Valley I want your father to write me something like the following—*

> *Dear Mr. Del Mar:*
> *I have Chief with me here at Cushenbury and will shortly be going to Bear Valley. You would oblige me by either yourself or Mr. Johnson coming over here and trade back the grey for your own horse. I cannot at present leave my family to visit you or I should bring Chief over to you as I know you would rather have your own horse back. This would be the most convenient time for me as I may go below sooner than I expected."*

The high desert region of Lavic is quite warm in the summer months, and Del Mar reports a temperature of 104 for that day. He also says he has assayed that morning and got poor results. *"It was mostly country rock,"* he writes, *"which contains a little silver and a trace of gold."*

News of the camp and its people is included here. *"Mr. J has sworn off (liquor) for a month. He has had a couple of gallons since he came up here but finding no one to help him empty the demijugs, he is getting tired of drinking alone."*

Later Del Mar adds that tells her Mr. J will be going to Bear Valley with him. *"If this is objectionable, let me know for in that case I might have to forego my visit."*

July 4 (added to above letter)

Perhaps to explain why Johnson will be coming to Big Bear with him, Del Mar writes, *"I am afraid I cannot very well desert Mr. Johnson. It would be a shabby trick so will have to content myself with a two day visit with Mr. J. He wants to see that part of the country."*

On the day the letter is written, Johnson has gone to Daggett for lumber and a windlass.

Perhaps in answer to a question she has asked, Del Mar says, *"My charges for assaying gold, silver, and copper will be $2.50 per sample, prepaid."*

Assuming all the equipment is in place, it would normally take about fifteen minutes to assay a sample.

July 13, 1901
Lavic, via Bagdad, California

Belle is getting a new dress, and she has sent a sample of the fabric plus a magazine with a picture of the dress. (Mrs. Rogers was an excellent seamstress who did sewing for other people as a way to supplement the family income.)

"Let me know as soon as you can when you will start for Bear Valley for we will start from here so as to meet you at Cushenbury. Mr. J. is very anxious to get Chief back as the grey is spoiling the other horse. Dutch [the other horse] *is much faster than the grey that as soon as he finds the grey is not doing his*

work he balks. It is no use whipping the grey for it is impossible for him to walk fast."

In an update on the work at the mine Del Mar writes, *"We are progressing fairly well here but finding nothing. About a month more of work will decide the matter."*

July 15, 1901
Newbury, California

The men have started toward Cushenbury sooner than expected.

"We are now on our way to Daggett, Cushenbury, etc. Johnson would not wait any longer to get another horse. The two horses do not and will not pull together and Dutch balks so much that it is almost impossible to get two barrels of water up to the camp. Sometimes they stick in one place for hours. I am sorry he would not visit longer as we will be using up all our time waiting for your folks instead of being with you. We will take in the whole county about the BH (Blackhawk) and Gold Mountain before you arrive."

Johnson's frustration with the horses makes his pressing to go on this trip more clear. At the same time, it explains why Del Mar bows to Johnson's insistence on being part of the trip, but it doesn't lessen the frustration Del Mar must have felt about spending their time waiting for the Rogers family to arrive instead of spending time with Belle.

More business enters this letter, with Johnson still making the decisions.

"Johnson says to tell your father that if he has a pair of heavy horses to trade for Dutch and Chief to bring them up if it is not too much trouble. As he will want to try them first your father can decide whether it is worth the candle. Also bring the collars for the large horses if he brings them."

July 19, 1901
Blackhawk Mine, California

Belle has evidently reported that a man named Harry Cahns has been paying attention to her. *"I don't blame him on thinking so well of you...if he thought otherwise I would feel like convincing him by force if necessary."*

Once again Del Mar gives us the route he has taken. *"We came across from Daggett through the Ord Mountains to Uncle Pete's to Cushenbury."*

He learned that Leach is back. *"They say Leach has been on good behavior for some time. He was pleased to see me and seemed in first rate metal. The garden in Cushenurry, I think, is better than last year's, especially the grape arbor. That is way up. You will want to get to Cushenbury for this alone."*

"It has been very hot this year Leach says. The springs at Cushenbury are not flowing half as much water as last year."

And then there is news of another old friend. *"Mac has just returned from the Desert and is completely broken up. He was sick for two weeks in Providence Mts."*

The travels continue. *"Tomorrow we go to Old Woman Springs and back to Cushenbury. Then we go to Cactus and over to Jim Smart's, then up to Holcomb and Bear Valley."*

He ends the letter with the hope that he will be able to stay at Cushenbury until she arrives.

August 12, 1901
The Stewart, San Bernardino, California

Dental problems have apparently forced Del Mar to seek help in San Bernardino. He has come by train in a ride he describes as *"long, hot, and dusty and...not at all enjoyable. Don't come this way if you don't want to get smothered in dust."* Del Mar had taken a stage that arrived at the Highlands in time to catch the 2:07 train to San Burdoo. He has recently been in Bear Valley and sends his thanks to her family, especially her mother, for a pleasant time.

"The dentist has already fixed my teeth (5 P.M.) as far as he can at present. The nerve has to be killed so I will have to return in about a month to have it filled. There were two teeth with holes and the one broken off must be capped." Sounds like a lot of fun!

Apparently Belle knew or knew of at least one other passenger who traveled with Del Mar, for he writes, *"The millionaire's daughter, as you call her, a Miss Hanson, sat on the front seat and did not eat any lunch at the station."* This letter also gives some insight into travel in the mountains and mine area in 1901. *"Five of the passengers were to come down by Walter Richardson but his stage broke down just before reaching Knight's Hotel and so his passengers were transferred. George Rathborn will no longer drive the stage but a brother of Chas. Henry will do so. It requires a good driver and a man of nerve to drive that stage down the steep grades, that is, to make any time. We came down in 8 hours, less an hour for dinner."*

August 12, 1901
A Second Letter from The Stewart Hotel,
San Bernardino, California

Del Mar writes a short note saying he has sent Belle a gift by Wells Fargo to Cushenbury via Victor to commemorate the anniversary of their engagement.

And he has apparently found something to soothe any pains left by the dentist as he *writes, "I have had an ice cream for you and am now going to have one for your mother."*

In note at the end he mentions cantaloupe is selling two for five cents and watermelons for a nickel each. Mining camps saw little fresh fruit and probably no ice cream, so Del Mar is enjoying these treats while he can.

August 15 1901
Camp Johnson, Lavic, California

He writes that he left San Burdoo Tuesday by train. *"At Newberry I saw Johnson with the team (2:30 p.m.) but train did not stop long enough for me to get off. I went on to Lavic where I caught a team going to the Wells. Mr. Shader asked me to supper and it was so poor that I was afraid he would ask me to breakfast had I slept there overnight so I went up the road a mile and slept, arriving in camp in time for breakfast."* Johnson arrived a few hours later.

The camp area had a big rain during which they caught thirty gallons of water, a fact Del Mar describes as *"quite a treat."*

Johnson apparently was carrying root beer. Del Mar comments that *"the root beer served Johnson well. His canteen leaked going over Ord Mt. and as he did not find any water on the way he used the root beer."* Johnson left Cushenbury at 7 a.m. and arrived at Daggett at 9 p.m. after traveling what he described as *"a terror of a road. He left Daggett Monday noon and had got beyond Newberry when he found he had lost his pocket-book. He went back and found it about two miles this side of Daggett."*

The strike at Bluebird doesn't amount to anything. *"The ore assays at 3 1/2 oz. silver and about 1 to 2 percent copper. It will require quite a lot of work to find a good ore body."*

Del Mar says they will give up there for the present after September 1 unless they strike something good by that date.

He must have sent Belle an autoharp as he explains how the music rack fits into it. He writes that she can see him at Cushenbury anytime the following week and that they plan to prospect a road to Gold Mt. the following week.

August 21,1901
Lavic, California

Belle is still at Bear Valley, where Del Mar assumes there has been a lot of rain as they have had floods of it in the desert.

"Johnson and I have been to Old Woman's Spring, up to the mine, and back to camp. We can go from here to Old Woman's Spring in about 12 hours walking the horses all the way. Only about six miles of the way is rough, the rest is easy ground for a good road. Johnson," he reports, *"is beginning to move camp."*

Del Mar urges Belle to try to get her father to go after the other mule as soon as possible as they are *"being greatly inconvenienced with the grey horse."* (This is likely the same grey horse that was such a problem earlier.)

"I will go over to the new mine about Sept. 1st I will then be about 20 miles from Cushenbury at that time. We must have a picnic and meet at Old Woman's Spring."

News of fresh fruit enters again as Del Mar reports bringing two watermelons back from the springs and says they ate two on the road and several more at Mr. Farmer's.

September 25, 1901
Gold Peak Mines, California

Mrs. Clark and her daughter have stopped by, and Del Mar learns from them that Belle and her family left Cushenbury on Monday, a day that was quite windy. *"It did blow great guns,"* he writes, *"and Tuesday it got quite cold."*

There is good news in this letter. *"We are striking rich ore at 30 feet in the shaft and think we have a mine."*

Johnson is still gone.

October 5, 1901
Gold Peak Mines, Victor, California

Del Mar talks about the irregularity of the mail, saying a regular route has not been established but after next month they will have their wagon going to Victor three times a month for mail and supplies.

More good news. *"We have a cook now and besides having better cooked food we have all our time in the mine."*

Later he notes the new cook's name is Johnson, making three men named Johnson in the camp.

"We are down 50 feet and have drifted about five feet. From 20 to 30 feet we went through some good ore 60 to 70 a ton and then we lost our ore and are now drifting to catch it." Drifting is a term the miners use to describe veering off the straight line of a tunnel.

To date, they have seen nothing of Belle's father.

The camp is apparently getting quite musical. *"Johnson* [which one isn't made clear] *and Clark are now scratching on the violin. I have had little time for the auto-harp and haven't therefore made much progress on it."*

In more mundane matters he notes he is sending five dollars for her to buy him a blue sweater and to use the rest for herself. He also thanks her for the clean pillow case and handkerchief she sent.

Mr. Farmer, the man who raised the melons Del Mar so enjoyed, is better, *"but still not strong enough to cut his hay so our cook when he went down for water on Tuesday cut it for him."*

The Blackhawk is still unsold: *"the people who were there* [the Bradshaw outfit] *will not take it, so all of Leach's vaporizing comes to nothing."* During this period Leach has been trying unsuccessfully to sell the Blackhawk. His "vaporizing" is his "talk."

October 14, 1901
Gold Peak Mine, Victor, San Bernardino Co.,
California

The mail continues to be erratic, causing occasional concerns for both Del Mar and Belle. In this letter he tells her that he is now having his mail directed to Uncle Pete's but gives no explanation as to who or what is Uncle Pete. Del Mar explains that the Old Woman's

Spring route has proved unreliable. Now, someone from his current team will go into Victor two or three times a month and collect mail and supplies.

Del Mar offers a glimpse into his future plans. *"I may go to Los Angeles next month but am not yet sure. I am trying to establish a business connection in Los Angeles which will keep me there most of the time. If I succeed in my project perhaps your school days will end again. Let up hope."*

This is the first mention of Belle's being a student again. It is unclear exactly what type of course she might be taking or exactly where she was taking it. Del Mar writes, *" I am glad that you are taking French and hope you will make good headway. It is a very nice accomplishment to have."*

Although Del Mar is one of six children he seldom seems to see his siblings. In this letter he does note that *"My brother Eugene was in Los Angeles when Lotus was there so she may have seen him. I was sorry he left before you and your mother arrived in town."*

In other news and comments he says, *"You are right about Mrs. Knight. She certainly holds herself very low but what can one expect. Her family certainly are not refined people, her brothers are saloon keepers. These are the class of people that pretend to be somebodies and are nobodies."*

Work at the mine continues and with some success as is indicated here: *"Johnson has been to Bear Valley again but did not have a break down* [in equipment] *this time. We now have two men working besides Johnson, the cook, and myself. It is my present intention to leave Johnson about the first of next year. I want to see him get a mine and by that time we will know. We have our ore down 50 feet so far and every indication points to the project being a great success. I am hiding myself too much here for I think my abilities are capable of being more useful in town than out here."*

November 8, 1901
Gold Peak Mines, California

Another indication that Belle is continuing her studies is revealed her as Del Mar congratulates her because her music teacher has chosen her to compete for a medal. There is no indication as to whether she is singing, playing a piano piece, or what.

Del Mar writes that he will be in town for Christmas.

Del Mar's partner, Johnson, is apparently involved in a law suit of some sort, for Del Mar comments that Mr. J. has no doubt been at the Rogers home by now and then adds, *"If he loses his lawsuit he will not be so flush with cash."*

News about the mine is good. *"The mine is still showing up well. We will soon begin to think about a mill to crush out the yellow boys."*

In another reference to Mr. J and money, Del Mar notes that when J. returns the two *"are to have a financial talk which will influence my future movement."*

There is news of people and places both Del Mar and Belle know. *"The Farmers have returned to Old Woman's Spring having been away for nearly three weeks during which time when Mrs. Clark and the girl took care of the place. Mr. Clark thinks of taking his family to Victor before Xmas time. He is now pretty sure of getting the $1250.00 due him on the bond. This comes due in February* *and will be quite a windfall. It is too bad the Blue Bird did not turn out well."*

November 21, 1901
Gold Peak Mines, California

Del Mar writes that lately the team has made quite a few trips to Rabbit Springs, affording him more chances to send letters to her. In this letter he details his plans for the Christmas holidays.

> *"My present programme in respect to Xmas is to be in San Burdoo on the 23rd to have my teeth filled, to be in Pasadena that night and stay until the night train on the 26th or 27th, yet to be determined."*

There is news about the mine and Mr. J who is still in town. J's *"suit has been postponed another week."* Later he adds, *"Mr. J does vex me sometimes. He is so unbusinesslike."*

In this letter Del Mar reports that at the mine they have a whim ready for a horse to do the work of hoisting up the rock.

The term *whim* as used here is a reference to a device commonly used in mining. A large capstan (a vertical drum around which ropes are placed) has one or more arms radiating from it to which a horse can be attached. The horse has the job of walking in a circle to cause the capstan to turn. One direction traveled by the horse causes the ropes on the capstan to unwind and lower a large bucket into the mine. When the horse goes in the opposite direction, the rope winds up and the bucket is brought to the surface.

Here Del Mar includes one of the less pleasant

aspects of mining. *"We killed 38 mice in four days besides have got as many more at odd times. Nig [the dog] caught two. He watches them just like a cat would."*

November 25, 1901
Gold Peak Mines, California

There is little news to tell in this short note. He is still waiting for a photo of herself she has promised him. Mr. J. still in town waiting for his case to come off.

Of interest is the passing comment *"Have you heard about the volcano struck in San Diego County while drilling for oil?"*

December 3, 1901
Gold Peak Mines, California

Del Mar writes, *"I may stay in town until the New Year if Mr. J can conveniently get along without me."*

The photos Belle sent have not arrived, and Del Mar suspects they have gone to Gold Mountain or some other place. *"The Turners at Victor,"* he writes, *"are managing the post office in a free and easy style."*

Del Mar appears to be feeling some frustration with both Mr. J. and Belle's father. *"Neither Mr. J. nor your father has arrived yet. As an example of Mr. J.'s free and easy style I have only received one letter from him since he has been away although I have written him many times and having asked several questions which I wanted him to answer. Your letter informs me that he thinks of leaving soon as well as your father but he has written nothing."*

In response to something Belle may have written, Del Mar tells her to definitely enjoy herself and not feel tied down by the engagement. *"You know you have my permission to go anywhere you wish, for I do not wish our engagement to interfere with your pleasures. As I am not able to be in town to take you about I want you to get all the pleasure you can without me."*

Belle has reported her weight as 112 pounds, a number about which she seems to have worried. The figure must represent a weight loss because Del Mar suggests she needs some more mountain air and adds that perhaps his not seeing her might contribute to the weight change.

Holidays in a mining camp often could not be observed, at least not in the traditional way. Belle must have asked him if he remembered what he had been thankful for at Thanksgiving, and he responds, *"Yes, I remember for what I was thankful for last year. I am the same this year. We had no turkey or even chicken at our camp and worked as usual."*

An update on the mine is included. *"Clark goes to town tomorrow to take his family to Victor. I will be the only one to work in the shaft until J. comes. Our shaft is down 60 feet—ore all the way—and is timbered all the way down so that it is perfectly safe. This is the second shaft—the other is 50 feet. We will run this second one down to 150 feet as soon as possible."*

Here the letters for 1901 end.

The Rogers family, minus George Rogers, about 1889.
L to R: Lotos, Belle, Rosa (Ma) holding Bruce, Reine

Camping at Big Bear Lake, early 1900s—George (partially
shown at far left) with Lotos, Reine, Bruce, Belle,
Ma Rosa Rogers and guest Algernon Del Mar

Belle and Algernon at Big Bear Lake early 1900s. Notice the tree stumps in the recently constructed and flooded Big Bear Lake

Cushenbury Springs Ranch, early 1900s—Lotos, Reine, and Belle Rogers. Notice part of the camp building behind the girls

Cushenbury Springs Ranch House around 1899

Black Hawk Gold Mine Camp with Del Mar family members, circa 1915

George Rogers harnessing donkeys to haul water tanks to
Black Hawk

Belle, Reine and Ma Rosa Rogers on the front porch of the
Rogers home at 2060 Alpha Street, South Pasadena. The Rogers
family moved from Pasadena to South Pasadena around 1910.
The families of both Belle and Reine settled in the same
neighborhood later

The Rogers family at the house on Mary Street, Pasadena, about 1902. L to R: Reine, Rosa and George Rogers, Belle, Bruce and Lotos

Algernon Del Mar, mining engineer, 1898

Belle Rogers Del Mar, circa 1904

Algernon Del Mar (second from left) with miners at Holocomb
Valley claim 1900

Walter Del Mar stands at the spot where "Cactus Jim" Johnston
had the Cactus Club on the west side of the present SR 18.
This is also the spot where the stage stopped before going to the
Rose Mine and Doble. The original building burned several
decades ago but the original road is still visible to the left.

1902

January 8, 1902
Victor, California

Del Mar writes that he has looked for Rogers at Victor but seen no sign of him. He is traveling with Mr. J and someone named Sam, but they have gone on *"to Daggett ostensibly on some business but as both were under the weather, especially J., I think Al Falconer's saloon is their goal. I kept clear of them while in San Burdoo and did not see them until the train came in. J. could just about stand."*

He reports he is writing everywhere for a new position as he is *"sick of this gang."*

After an update on his dental work he adds, *"Victor is as slow as ever if not slower."*

January 11, 1902
Victor, California

Rogers arrived the previous night, and they expect to be on the way soon now. More complaints follow about Johnson. *"J. has spent all his time in Al Coffman's saloon, leaving me to think of everything. I have set the wires* [telegraph] *going as hard as I can for another position and will not await development."*

Del Mar must have been feeling a lot of dissatisfaction with his present position. Not only is Johnson irritating him greatly with his lack of paying attention to business, now Del Mar tells us the following about a man named Clark who has some position of authority at Gold Peak: *"Clark's dislike of me is simply jealousy. He does not take into consideration that if it was not for me he would still have his prospect. I'll get even with them in time."*

Alexander Del Mar, Algernon's father, has written that *"the Black Hawk is nearer being sold than ever, that papers for a short option would be signed Dec. 31, 1901. A telegram has been sent up to Mac[Fee] and Leach to ship ten tons of ore to New York so I suppose the papers were signed."* It is likely that MacFee and Leach owned owned most of the Black Hawk and that Alexander Del Mar had a one-quarter interest in it.

January 18, 1902
Bear Lake, California

Belle has questioned Del Mar's saying he would not await developments. He meant to say he would wait developments.

"I am up here with Dave Welch taking down the old mill. It is to my interest to stay about here to what will be done at the B.H. so I might as well be working." He is apparently working at Gold Mountain.

"From Victor the gang went to the mine (at Gold Peak). J. was the rockiest of them all. Dave Welch, Johnny Rosson, and myself had not been drinking

nor had your father who came along with J. We stayed a day at the mine then your father left for other parts. Johnny Rosson and Al Parr were left to dig a well. J., Welch and myself got to Cushenbury, arriving around 5 p.m. and rather than stay with that old fool I walked over to the B.H. to see Mac. The next morning I met the team at Cactus and came up here."

More complaints follow about Johnson and his behavior. *"Johnny Rossan and Dave Welch were disgusted with the way J. managed things at the mine. I said nothing nor did I offer advice. Everything was upside down, men doing nothing, team not ready although there are two teamsters. Johnny says he does not want to work up there as there are two bosses."*

The weather, he reports, is cold, and there is a possible storm brewing.

January 23, 1902
Bear Valley, California

Del Mar reports that they will be finished there in a day, and he is glad, since the weather has turned bitterly cold.

"About 1/4 mile from us is a warm spring where the ducks sleep at night. Having no gun with us we have been compelled to go without duck although someone goes to the pond and shoots the ducks every morning."

Two snowstorms so far have made life less than perfect in the mountains.

Although there did not seem to be a gun with

which to shoot the ducks, the camp is not entirely without weapons. *"My dog 'Ninkey' (otherwise known as Mace alias Nig) is a fine bird dog as I have found out. I take him over the lake to practice although I use my revolver yet he acts well creeping close behind me. All I have to do is point my finger at him and he stands still as a rock. But when he hears the gun go off he is all over the place looking for the game which I did not shoot."*

He reports more dissatisfaction with those around him. *"Leach is an old fool. He will get something near $10,000 if the B.H. is sold. One glorious booze and no more Leach. He was drunk the other day as I passed up."*

Part of Del Mar's criticism apparently stems from a fear that Leach will continue to be drunk and will lose the money.

Del Mar speaks here of going back to the mine by way of Cushenbury. Another option is to return to the B.H. and meet the team on the other side of Cushenbury. At any rate he wants *"to see as little as possible of that ungrateful hypocrite."* The reference seems to be toward Leach.

January 27, 1902
Cactus Flat (Doble P.O.), California

Heavy snow has fallen with six new inches. Del Mar writes that he and Dave have finished taking apart the mill in Holcomb Valley. The snow arrived before the team came, so rather than get snowed in the men walked to Gold Mt. From there Del Mar continued on to Cactus, where he knew he *"would find a hearty welcome and a good fire. I did not wish to bum around Gold Mt. sitting about the stores."*

Bruce (Belle's brother?) has arrived but is in town to see a specialist about an ear problem.

Del Mar takes advantage of the free time to visit Mac, whom he reports has still has not fully recovered from his sunbath the previous summer. Remember, Mac had been out on the desert alone, became ill, and was not in very good shape when he returned.

Another friend, Johnston, has made some of his "famous" pie for Del Mar and had instructed him to tell Belle that he (Johnston) was feeding Del Mar on his fine pie.

"The Black Hawk affair has progressed no further. When the 10 tons of ore reach New York the parties have a two month option."

Blackhawk Mt. is covered in snow, and some has drifted to over a foot deep. Johnston *"has regaled me with stories of the Bannings, Bradburys, Nickols, etc. etc. families. He has nearly talked me deaf and dumb. They say at Victor that Mrs. Clark is tongue-tied (that is a Victor joke for she is really a talking machine). Well Mr. J. of Cactus would run her a close second."*

As for the Gold Peak mine, the news is not good. *"It appears that Mr. J. of Gold Peak is getting nearer and nearer to making a failure. He has consulted me but little lately and is rapidly throwing his finances away."*

And he continues, *"He (J.) started digging a well in a place where there is no indication at all for water and considering that the success of the undertaking makes it imperative for him to get to crushing ore as soon as possible it was surely his policy to go where he was sure of water. By his method of delay (the result of booze) part of the mill is snowed in at Bear Valley and Dave and I are idle here. I am being paid by the month and will draw my pay or have a row with him should he want to pay only for the time I was working. As for Dave, he can look out for himself."*

On a more positive note he writes that he has received a nice letter from Senator Jones, enclosing an introduction to the manager at the California King Mines. Del Mar says he has *" forwarded this and my testimonials to Picacho [?] and if anything comes of that he will immediately leave Whiskey J."*

In a postscript that seems to respond to a letter he received before he could send the part above, Del Mar continues his criticism of Leach. *"Mr. Leach will get little of my company. I have guessed all along that he has been writing things and I think so now. If I should repeat what I hear that he says when he is drunk none of you would stay there a minute. He is an undisciplined old cur and deserves no consideration whatever."*

Apparently there has been some gossip about Del Mar and Belle, and she is concerned. He answers her, saying, *"You must not mind what you hear respecting me or your actions and always if people should write you it is best and right to show me the letter. Many people are always on the alert to put their say in and from jealousy or just bad nature are always trying to make others unhappy. This subject is one I wish you would discuss with your mother for it is one where she can give you much better advice that you will always be glad of remembering."*

February 10, 1902
Gold Peak Mill, California

This letter begins with telling that both Mr. J. of Cactus Flat and J. E. MacFee wish to be remembered to Belle.

He also sends regrets that he is not in town to take Belle to see the Shakespearean plays that are being presented there.

After sympathizing with an accident of some sort that Belle had and in which she was not seriously injured, Del Mar tells of a somewhat humorous mishap he had while in Victor.

> *"Jimmy Curtner was about to close his shop for the night and I went into the back room to get my coat which I had left there. It was pitch dark so you can imagine my surprise which I picked myself up in the cellar eight feet below. He always opens the trap door so that the cat can get into the cellar and I had walked right into the hole. I only sustained a slight bruise on my skin but it might have killed me. I thought of you as soon as I touched ground and when I found myself unhurt, I thanked God for so narrow an escape. Although it was only eight feet but it seemed a long time and I thought a great deal in that time."*

More work delays are recorded here. *"The mill stuff has not all come down from Gold Mt. The roads are too slippery and there is too much snow to be able to get everything out just now."*

Del Mar writes that he is sending Belle his insurance policy and instructs her to keep it in a safe place. He is also having sent to her an assignment, which, once she fills it out, will assign the policy to her.

On a negative note, Del Mar writes that water has not yet been struck at their present location.

The following statements probably refer to Mr. J. *"I*

like your new name for a certain party 'the-one-horse-whiskey-outfit.' It fits nicely. Last trip he left Victor with a gallon demijohn, a half gallon demi and a quart bottle. The boys at the mine must have had a good time. The next day Al Carr came down for water and said he (Carr) had been pretty well overloaded the night before. It is a shame and his resources are fast fading away."

The men in these mining and mill sites did apparently try to stay abreast of the news. Del Mar writes, *"By the way thanks for Sunday Times received. They came in handy to face the evenings. Johnny Rosson gets the Weekly Examiner so we are fairly well fixed for literature."*

February 17, 1902
Gold Peak Mill, California

A large part of this letter wonders if Belle has received a photo he sent of himself and a lament that the photo is not as flattering as he hoped it would be since it was taken by a professional.

A book the two have previously discussed is now back with Del Mar. *"Last time J. went to Gold Mt. I had him (J.) ask Leach for the book Middle Ages which I now have here at the camp."*

They are enjoying spring weather, and the wind storms so common at this time of year are absent.

Regarding the Blackhawk, *"I do not think Black Hawk matters have advanced any further. Mac and Leach have not yet shipped the ore. They are dead slow. Mac is still troubled by the effects of his sun bath last year and wants to get away to those hot springs near Warner's Ranch."*

What is referred to here as Warner's Ranch is now

Warner Hot Springs, about fifty miles from the Black Hawk by way of San Bernardino. Warner Ranch had served as a stage stop and a spot where travelers could rest and find food. The early travelers also learned that the hot springs afforded possible therapeutic benefits to those who bathed in their water. It is still a popular retreat located on Highway 79 in San Diego County. Today, though, several hundred cabins and several restaurants have been added to the property.

Del Mar reports that *"Johnson is drinking as much as ever if not more"* and that the *"bobtail-whiskey mine is still running."*

Finally, he adds, *"When your father went away to Manvel he left a club foot lame horse at the mine to work the whim. Johnson says he eats a bale of hay a day and a bathtub of water for each meal."*

February 24, 1902
Gold Peak Mill, California

In this letter Del Mar reports, *"The mill is about half up now and the worst part is over. On account of the well diggers leaving us, nothing has been done on the well for over a week so we do not know when water will be struck. I am expecting two men tomorrow. I am in a hurry as the team will go in tomorrow and is now on its way from Old Woman's* [Spring] *with water for the mine."*

Del Mar has been advertising in *Engineering and Mining Journal* for a new position. He has now received an offer for a superintendent at a mining company in Nelson, B.C., and has answered it. *"Tell me,"* he writes, *"shall I go that far. If I do you will have to come with me later on as soon as I can see what sort of place it is. I am*

101

looking forward to going to Bear Valley next summer to stay with you for two months."

February 24, 1902 (second letter of this date)
Gold Peak Mill, California

Del Mar needs copies of several testimonials (letters of recommendation) to submit to possible employers. In the days before copy machines, getting those copies was a slower process. Del Mar here instructs Belle to go to a typewriter (person who types) to get copies made of the eight testimonials he is enclosing. He adds the name of the typewriter he wants her to visit as she has made copies of these letters before and will know how he wants them. He estimates the cost will be about $1.50.

In this letter he also sends an address for *The Mining and Scientific Press,* and asks Belle to subscribe to it in her name. He encloses a money order for $15, which he explains as $9 for the subscription, $3 for the typist, and $3 for envelope. Anything left over she is to use as pocket money.

"J. is now at Virginia Dale looking after some cyanide tanks and when he returns I will ask him to settle my month's wages with me and give him no credit. Also as I promised him a month's notice before quitting I will wait until March 15 and give him notice to quit on April 15. This I have decided to do for two reasons—first is to the first proposition I will take your advice which is strengthened by the second reason, namely that whiskey by the bottle is not enough at the mine but they are getting cases of it. Every time J. goes from Victor or Gold Mt. to the

mine he has a couple of demijohns. Now if he can afford to throw money away when he needs it so bad I am not going to remain here and bear the burden for as sure as white is white, I will get blamed for a failure and not the whiskey."

He reports that Dave (Welch) is laid up with a bad jaw—Del Mar and Dave were lifting a heavy casting when *"the crow bar slipped, hitting him in the jaw and me on the toes. Fortunately my thick shoes deadened the blow and it did not hurt me much but Dave got the full force. His face very much swollen."*

As for good news, Del Mar says the ore from the Black Hawk is now being shipped.

March 10, 1902
Gold Peak Mill, Victor

"We struck water here at 63 feet, good and plenty of it." This water supply will make both life and work at the mill easier. We are not told if this water was located at the place J. designated, which Del Mar thought was not a good spot, or if the well ended up in another location.

In another new development, Del Mar writes, *"I have decided not to leave here until July unless I get another place first. If the British Columbia and Colorado parties want me for sure than I will give a month's notice from the time these people want me. This will be better than leaving on April 15th without anything in sight. However anxious I may be to see you and be near you yet I hate to be knocking around Los Angeles doing nothing. What do you think."*

"J. was at Victor three days ago but went back to Los Angeles to get something that was forgotten. He will be overjoyed when he knows we have struck water."

March 17, 1902
Gold Peak Mill

This letter tells Belle that Del Mar has given J. notice that he is quitting *"as soon as it is convenient to him."* Further, he has asked J. to write him a letter of recommendation and to send or give it to Belle. He then instructs Belle that if this happens, she should go immediately to get four typewritten copies and send them to him while keeping the original. All this activity, he says, is part of his little game, which he will explain some other time. (The explanation either never came or was delivered orally. It is not in the letters.)

March 18, 1902
Gold Peak Mill, Victor, California

Most of Del Mar's family has remained in New York. They are aware of his engagement to Belle, and Del Mar's mother has written to Belle in the past as a way of welcoming her. Now Del Mar tells Belle, *"My people at home are continually asking me when our event will take place and if I am coming to New York for our honeymoon. I will be satisfied to get a good position in a pleasant locality and let New York go until we could go there comfortably and at leisure."*

The British Columbia man wants to see Del Mar in San Francisco, but the time is very indefinite. Del Mar has written the man that he can not leave the mill on an uncertainty. He suggests the man will need to come there.

Work at the mill is progressing. *"Our mill is nearly completed. We are only waiting for the engine and to put on finishing touches. I think we will start going about April 1st. Somebody may get fooled on that day."* There is no explanation given for that last statement, but it could be a reference to Del Mar's possibly leaving and J. would be the one fooled.

Next he praises Belle for going into the pie business and writes, *"I have a splendid mouth for pie, just made for pie, sort of twitches every time I think of pie, and if your pies are as good as your mother's, which I am inclined to think they are and perhaps better, why we will live on love and pie. How is that for sentiment, five cent novel style, sort of piebald sentiment is it not."*

He plans to go to Bear Valley the following week to get some stuff at the old mill. He plans to see all their old friends on the road. *"Leach sent me a bill for about $6 for meals I had there last July and August. Now whether this is when I was with you folks or not I will find out. The old hypocrite is up to some game. I put my father on guard against him so perhaps he is trying to rake up old cases against me and this is all he could find. I think, however, it is for meals I had while doing Johnson's business and if it is J. will have to pay."*

Religion and going to church have never been mentioned in the letters until now. Del Mar writes, *"The Brown Books you sent came in very handy. Did you read an editorial about going to church for if you did you will have my argument but in more fluent language than I can command. If not I should like to send the article to you."* The article is not available, so we are left to wonder just what it said as well as Del Mar's thoughts on the subject.

He adds here that there are three feet of snow in Holcomb Valley so the lake will be full next year.

Newspapers then, as now, sometimes ran contests to garner new subscribers. In this letter Del Mar reports, *"I have subscribed to the Weekly Examiner and, of course, expect to get the $10,000 prize. My number is 84412. Quite a pretty number."* However, there is never a mention suggesting Del Mar won that prize. Moreover, he tells Belle to definitely continue sending the *Times* as they would all miss having the reading material.

In one portion of the letter Del Mar speaks of looking forward to spending time with her in Bear Valley in the summer, then speculating that he might need to accept an offer in British Columbia or Colorado. He concludes by saying that no matter where he is next year, he wants her with him.

"Johnson came out from Los Angeles," he writes, *" and looked a perfect wreck, face swollen and voice husky."*

A small note on his family is included here as he tells that his sisters (in New York) had the first sleigh ride they can remember ever having, and they loved it. Del Mar notes that the opportunity for a sleigh rides in New York are infrequent, but *"in this land of sunshine, wind, and cactus not at all* [are sleigh rides available], *although we have had the thermometer low enough."*

March 23, 1902
Gold Peak Mill

"I have my programme now made out for Mr. J. and will fire off my heavy artillery in the near future. If he has written for me the letter I asked for I will demand my pay in full and that [the letter] *not forthcoming, I will threaten him with a lawsuit for 1/10 commission on the mine or $500 due for ser-*

vices as per our agreement with Robinson and also put an attachment on the mine for salary. If he has not written the letter I will carry out the same plan, as I then will know he will give me no letter. I can not for the life of me see how he can make the mine pay the way affairs are now conducted at the mine."

Del Mar, as we have seen, is not happy with his present job. He has continually written about J. and his (J.'s) fondness for whiskey and the way he seems to pay no attention to business. To a professional man like Del Mar, these were serious weaknesses. He continues his complaints here with the following: *"There is no head and everyone does as he pleases and important work that should have been done has been entirely neglected so that the mine is in no condition for taking out ore."*

He tells Belle he expects to be in Los Angeles about April 20th, *"position or no position."*

George Rogers has been at the mill and left some ore to be examined. Del Mar has examined and now asks Belle to tell her father that the ore is of little account, although it does indicate that some gold is at the spot where Rogers got the ore.

Here Del Mar tells more exactly where he is, writing that the mill is *"about six miles north of Old Woman's Spring about 14 miles east of Rabbit Springs and two miles south of Copper Peak. It is on the western edge of a dry lake and plenty of water at 63 feet."* Then he adds this information: *"We now have the engine and other material will come later. We will not have it running until some time in April."*

Almost in passing, Del Mar notes that her father will probably tell her about the Vader runaway, which

Alg calls disgraceful. There is no further explanation of the matter.

> *"Have heard nothing about Black Hawk except that the shipment of ore was on its way to New York and that the sale depended upon how economical it could be worked. I fear the result will not be satisfactory for technical reasons which I need not go into."*

Del Mar has just told us in this letter that he is six miles from Old Woman's Springs. Now he says, *"I might walk to Old Woman's tomorrow to tell Mr. Farmer there are some boxes here for him and to get a few lengths of pipe."* Much of those six miles is relatively smooth, but there is still a mountain to climb to get to the mill.

March 31, 1902
Gold Peak Mill, Victor, California

Belle has sent him a box of chocolate creams, for which he thanks her. He could, he tells her, only have sent her some sage and greasewood (which he didn't do).

The weather, he reports here, sounds miserable. It has been very stormy with cold winds, some rain and snow.

Belle has apparently written about a proposed dance hall in her area, and Del Mar responds, *"So Harry Myers is putting up a dance hall, I suppose for the bashers. You are right, no self-respecting girl would go there if it is as I presume."* The specific town where this is to be built is not mentioned.

Del Mar asks her to post an important letter for him. It is an answer to an ad in the *Mining Journal*, and

Belle is to have certified copies of his testimonials made to send with the letter. Then he tells her he is enclosing two letters he has received that contain possible job offers. *"What should I do?"* he asks. *"Accept the first offer* [the one from British Columbia] *or wait for the Colorado business. I think the former plan best. What do you think or prefer me to do?"*

Returning to life and work at his present position, *"We have the mill nearly finished and only need the engine, etc. We have not struck water yet. We have a camp of four— Dave, a miner, a cook (and windlass man) and myself. The mine is looking first rate, so I hear, and improving as they go down."*

The mention of not striking water written here is at odds with information in other letters. On March 23 he says (not for the first time) that there is plenty of water at sixty-three feet.

April 1, 1902
Cactus Flat, California

After assuring Belle that this is not an April Fools letter, despite being written on April 1, Del Mar says he has not had his talk with Johnson so doesn't know if his programme will be changed. Del Mar has apparently received a letter from Johnson, quite likely the testimonial he had requested. He calls getting the *letter "a good stroke for now I have it in his own handwriting just exactly what I did for him and he can only perjure himself by running me down."*

The weather is improving a bit in Bear Valley, he writes. Most of the snow is gone, and there is a lot of water in the upper lake (Baldwin Lake).

On his way to Cactus, Del Mar passed Cushenbury. He only said howdy to the proprietor, who *"looked a little seedy,"* but gave Del Mar *"his hypocritical hand to shake."*

The British Columbia man has written, and Del Mar thinks he must have given the job to someone else since he was unable to go there to see him. There have been no responses yet on jobs in Colorado or from Piecho (but he did get an offer from a Minneapolis group offering a superintendancy job at a copper mine in Utah). *"If nothing else turns up, I will bravely put my shingle up in Los Angeles and await events."*

George Rogers keeps hoping to find a gold strike and has again sent ore or information for Del Mar to analyze. Again, Del Mar writes that the hoped for riches do not seem to be in Rogers' future. *"To tell the truth I took very little stock in your father finding what he was after but he was so enthusiastic that it was almost comical. He is the thousandth one that has tried and failed and so they go. Perhaps he will find it next time."*

Mr. Johnston, he says, sends regards.

April 3, 1902
Gold Peak Mill

Del Mar is now back at camp but did not get to talk with Leach, since he was not at Cushenbury when Del Mar was there. Del Mar has, however, talked with Johnson, and *"he agrees to pay A $500 for my services. The money* [is] *to come out of profits, including 5% of net profits until this amount is paid. Under these circumstances I could not well refuse to stay until June 15."*

He now plans to pay a visit to Pasadena in about a week, where he will stay a couple of days and then go back to the camp until Belle's family is ready to leave for Bear Valley. At that point he will leave Johnson but has not promised to return. *"We will have a great time in Bear Valley this summer."*

April (no day), 1902
Rabbit Spring, California

In this letter we learn Del Mar has been to Pasadena and has not left without some affliction. He has a swollen eyelid, which is causing him some concern. He asks Belle to ask her father to inquire about it the next time he goes to his specialist and send whatever medicine is recommended. *"The symptoms,"* he writes, *"are simply these—the right upper eyelid is swollen, especially near the lower portion and tends to keep down there by closing the eye. The eyeball and lower lid are all right as far as I can see."* Then he adds, *"I have a cold and a slight sore throat and have bandaged up the eye with a milk and bread poultice and will do no reading."*

April 18 1902
Gold Peak Mill, California

His eye is now fine, and he thinks perhaps he had measles and was hasty in describing the eye problem. *"Perhaps I was over hasty in writing to you from Rabbit Spring but one can not tell, sometimes small things lead to big results and having your father's experience fresh in mind I thought it best to have him send some medicine."* There is no explanation of what that recent experience was but written as it was suggests George Rogers failed to treat an illness or condition until it became a big problem.

Work goes on, and Del Mar expects the mill to be going in a week.

April 24, 1902
Gold Peak Mill, Victor, California

There has been no recent word from Belle, and he is concerned she may be ill. For his part, he tells her his eye is no longer bothering him and he will not need the powder from her father.

> *"We are now sinking our well to get a pump of water. The engine works first rate and after our well is sunk, we will begin to mill ore. There is too much copper in the ore I think to cyanide it and we will just have to run it over plates."*

Del Mar writes that expects to see Belle's father out there in a week or two and hopes he is *"over the measles and sundry complaints."* He stills wonders if his own eye problem was related to the measles, even though he did not ever break out with the measles.

Finally he writes that he has not found his dog yet, *"or more properly speaking, he has not found me."* He has promised the dog to Mrs. Farmer because *"I do not care for a dog that will leave me to follow a lunatic."*

He never does say who is the lunatic that the dog followed, but one can surmise it might be Mr. J.

April 28, 1902
Gold Peak Mill, California

Life, we learn in this next letter, has dealt a very sad blow to the Rogers family and, by extension, to Del Mar. He has just received a letter from Belle's father and her Aunt Belle informing him that Belle's younger sister Lotos has died. The reader will recall that Lotos and Belle were very close to each other and realize that this death would be especially hard for her to accept.

"The news has quite staggered me," he writes, and then he counsels Belle, *"Don't grieve my love, keep up courage and try to remember that she is in a better place now where all cares and troubles cease. In that far off place where God rules, all is love and happiness and there dear Lotus is now. It is like losing my own little sister and I am not ashamed to say that I have shed tears of grief and sorrow for her dear memory."*

Lotos, then nineteen, died of scarlet fever, a disease that in 1902 was all too common despite efforts to quarantine victims and all who were exposed to the disease. Now Del Mar cautions Belle to take care of herself and wishes very much he were there to comfort her. However, he points out it would do no good for him to come to Pasadena, as she is quarantined. In fact, Belle cannot even write him now, as there is a ban on mail coming from a place that is quarantined.

The following points out how short money is in the Rogers home. Del Mar writes, *"Rest at ease, my love, about the burial expenses which your kind aunt has so delicately mentioned. I will see that they are paid and will consider it a slight mark of my affection for the memory of our beloved sister Lotos."*

"When all this [quarantine] *is over, of course, you will want to leave town and enjoy quiet,"* he writes. Since it will be too cold in June in Bear Valley, Del Mar suggests the family come to Old Woman's Spring for June. *"Here you will enjoy quiet and Mrs. Farmer is good and kind and*

I will be able to see you all...Oh Belle, this sad bereavement seems to have brought all the love that is in me to you and I will pray the good Lord to spare you for me, for without your loving heart life would be a blank for me. If you should be in danger send for me and I will come."

May 6, 1902
Gold Peak Mine

Since there has been no word from Belle he assumes she is still imprisoned (quarantined).

"We have the mill running," he writes. *"It isn't going as smoothly as anticipated but we will fix that in time. It is very dusty in the mill and I hope we will take out enough gold to buy a good stamp mill some day. It is uphill work using old machinery but this was the best that could be done under the circumstances."*

Del Mar reminds her that he has no way of sending letters where he is and asks her not to worry when she does not hear from him. He hopes her father will drop him a line on her condition.

More names of those working at the mill are also mentioned when he writes, *"Frank Preciado is hauling down ore for us and Al Carr is helping at the mill."*

There are several signs of spring beginning to appear, Del Mar notes. *"Birds are beginning to come to the camp for there is a running stream of water all day from the plates."*

May 12, 1902
Gold Peak Mill, Victor, California

Belle is still quarantined when this letter is written. His attempts to send an earlier letter failed. What follows illustrates the lengths to which Del Mar had to go for something as simple as mailing a letter.

"There are three young men," he reports, *"have taken a bond on a prospect near the Gold Peak and as they were going to Victor on Tuesday, I walked from here to intercept them but got on the wrong road and had the chagrin to see them on another road but too far away to be able to attract their attention, so I had twelve miles walk before breakfast for nothing. It was surely 'Love's Labor Lost.'"*

He has heard the quarantine is lifted and the family is safe. He sympathizes with Mrs. Rogers and Belle for the predicament *"that fate has played you. Bear up, my dear, and try to look at the bright side of the picture. I know there is not much consolation in this but try for the sake of your own health and my happiness."*

Del Mar also asks Belle to send the name of the undertaker who was in charge of Lotos's remains. (Remember, Del Mar has earlier said he would pay the expenses for her burial.)

Del Mar is quite anxious to see Belle and be certain that she is healthy. Now he refreshes her memory that in his last letter he had suggested that if she and her mother wished quiet in which to rest that they come to Old Woman's. *"Mrs. Clark and her girl will be out there when school is out so that if you should decide to come for a month you will not be entirely alone. You could get your mail and provisions through our camp."*

May 17, 1902
Gold Peak Mill, California

Del Mar's relief at once again hearing from Belle is obvious in this letter. He writes, in fact, that he is happy to hear that all of them are recovering from the recent sad event and he is so glad she can now send letters to him. *"My heart is with you, my dear, in your sorrow and anything I can do to lighten your burdens or that will give you a particle of pleasure, I will do it with a glad heart."*

"You were right dear," he continues. *"I want to be more than a friend to you, I want to be a help and support to you in your hours of need and sorrow and never fail to call upon me for I will respond with all my heart and soul. A man that has his first real love at my age has it for good."*

Mrs. Rogers has apparently said "no" to Del Mar's suggestion of camping at Old Woman's, for Del Mar writes, *"Of course I am sorry you will go to Cushenbury for I will not be able to go that far to see you but I could get over to see you at Old Woman's after the day's work. But as your mother says, the welfare of her children stands first. You [Belle's family] must manage to come over and see me and perhaps camp a night or two at our well. Bruce* [Belle's brother, now about 5] *would like to see the "shenery" and he [Alg] would like to see her."*

Finally, back to life at the camp, Del Mar says J. has just returned from Victor with a comfortable "jag" and three more whiskeyites.

May 20, 1902
Gold Peak Mine, Victor, California

This letter is to Belle's mother. (Her letter to Del Mar was enclosed in Belle's last letter.) *"Words cannot express my sympathy with you and yours in the sorrow that has befallen,"* he writes, *"nor can I ever repay the care and watchfulness that you have taken of our own dear girl."*

"Under the circumstances," he continues, *"I think your*

choice of Cushenbury a good one, although I should like to see Belle more often. You might come over here for a day or two and camp near our well."

Then he asks her to pass the following news to Belle: *"Al Carr's uncle has died and left him $62,000. He (Al) may go in with Johnson on the mine. 'Sky' [whiskey] will surely flow."*

May 31, 1902
Gold Peak Mill, California

In this letter Del Mar sends a check for $135. He asks Belle to cash it and take the money to Adams Turner (the funeral parlor).

> *"I hope to see you soon out here. I have asked J. to get a man to take my place as soon as possible as I do not want to stay here. As I anticipated, J's bad management he is trying to shift on to me. He is running into debt all the time worse and worse."*

Del Mar is glad he did not answer the account Leach sent him and thinks if he can settle it up, the proprietor at Cushenbury will welcome him. This is a reference to the bill for $6 Leach sent to Del Mar for meals eaten.

June 7, 1902
Cushenbury, California

"Luck still seems to be against me," he begins. *"Here I have come from the mill to see you and lo you are not here yet. I came over yesterday morning and must go back today as the team [which he has driven over] must take water to the camp tomorrow."*

He continues by saying that he expects to leave the mill and the whole outfit for good as soon as J returns or as soon as J. gets a man to take his (Del Mar's) place.

"I will come directly back to Cushenbury with my goods and chattel or if for some unforeseen reason but I decide to stay awhile I will let you know as soon as possible. If I do not come over [to Cushenbury] *by next Sunday week, could you not manage to come over with your father and the whole outfit for a day. It takes four hours (horse walking) from the mill to Cushenbury on the road straight across the desert or perhaps your pa might manage to ride over and make arrangements for me to meet you all at Old Woman's Spring. I think, though, that you will see me over there on Sunday even if I have to walk. If no one comes to take my place, I will stay a while longer."*

August 23, 1902
Victor, California

Despite high hopes of being paid what was owed him, Del Mar has yet to receive the money. We are not sure exactly when he was last at the mine. Since Belle was nearby at Bear Lake during much of the summer, there are no letters to tell us some of the facts. In this Aug. 23, 1902, letter, however, we learn some of what is happening.

"Johnson has let his case go by default and I have a judgment against him. The bullion will be sold next Friday at 10 a.m. If your father comes before then I may try to persuade him to prospect for a couple of days and if he does not

appear until then I may go back to Bear Valley without him if nothing turns up."

To date, he has had no answers to his inquiries at other mines.

"I hope that you will not be lonely and that you have made friends with the Reverend gentleman and his family." There is no clue as to what that statement is about, but it might be a reference to someone in Pasadena.

> *"Victor is very dull, most everyone being away. It is a torment to be here with nothing to do. As it was I need not have come down but I did the safe thing for had J. come out here with a bluff he could easily have got the bullion. I have the law now from Judge McNutt and no bluffs will go."*

In closing, Del Mar notes that a Mr. Martin has been nominated for justice of the peace (*"what a travesty of justice!"*) and One Lung Brown for constable.

August 26, 1902
Victor, California

Del Mar writes that he is very tired of being in Victor and blames J. for his having to stick around there. To date, George Rogers has not appeared either, and Del Mar thinks now he may go with Rogers when he does show up.

> *"Johnson has made several bluffs but so far of no avail. I am not over confident however, but will know on Friday morning when the sale should take place. Last night his lawyer telegraphed to stay the judgment but the Judge had already given judgment.*

*He (J.) is very petty in thus trying to cause trouble.
All he can do now as far as I can see is appeal and
put up bonds for twice the judgment and a hundred
dollars besides, or in all $606. This I hardly think
he is able or willing to do."*

Del Mar speculates that if he can buy some of the
bullion for less than it is worth, he can in that way
make J pay for some of his lost time. *"I must acknowl-
edge that the 'book test' not answering accurately to who
should get the bullion causes me some misgivings and I will
not feel at ease until it is all over. He has ten days in which
to appeal or until the 3rd of October."*

*"So far the book has predicted correctly and I hope the
future will have something good in store for me."* The book
test is probably a reference to what is, or is not, stated
in the law books at hand.

Periodically these letters have revealed various
injuries that Del Mar or others at the mines or mills
have encountered. There has been little mention of
medical care, but now we find that at least in some
places there is a health plan of sorts. Del Mar describes
the situation at Gold Mountain:

*"I hear that there the men at Gold Mountain feel
very sore about being compelled to pay $1.50 or
$1.25 a month for doctor's fees besides which they
do not like the doctor. I am told the men are ready to
quit work and leave at any time if they can only dis-
pose of their dwellings. They are complaining of
being worked too hard."*

"I suppose," he continues, *"Dr. imagined he was cut-*

ting quite a deal with the Company's team. Perhaps he now owns the Company."

Del Mar writes he had no luck on a recent deer hunting trip. He tells of hearing that seven men from Gold Mountain went up behind Mr. Knight's place for deer after someone reported a herd near there. No word on what sort of luck those men had.

Please remember that it is well over a year from the time Del Mar had originally planned to marry Belle. It is two years since they became formally engaged. That time lapse and all that has happened—job disappointments, Lotos's death, Belle's own health—have not, however, diminished Del Mar's determination to marry the woman he loves.

"Well, don't be surprised to see me up with you soon. I am pleased that your love is getting strong enough for you to go with me when my affairs look brighter than they are now. I know you would be as happy and I dare to think more happy than you are now. I know I need sympathy and encouragement and with your sweet help I think I can make a better fight than I have done so far. My nature needs its counterpart to round out my life to its full extent. I am like a twin screw steamer with one screw disabled. I hope my affairs will take on a better shape and fortune and my experience will help me in the future. I love you dear and always will, no matter what befalls."

While the twin screw simile might not be the most romantic ever written, it does reveal Del Mar's feeling. And please remember, he was an engineer so to him this was a perfectly natural comparison for him to make.

There are no letters to tell us the final outcome of the suit against Johnson. Nor do we know why in the following letter Del Mar is traveling to San Francisco by boat rather than by train. It is, of course, possible he has been to British Columbia to investigate some job offers. Or he could have been in the Los Angeles area and thought this would be a quick and inexpensive mode of travel.

December 30, 1902
321 Post St., San Francisco, California

"Arrived here safely this morning. The trip was not very pleasant, the boat being small and with no accommodations. It was one of the lumber boats going back empty. There were some 17 passengers all more or less disappointed. We struck quite a rough sea which made the little boat pitch and toss."

He writes that little has changed since his last visit to San Francisco. He has looked up a few friends and found lodging at the Post St. address listed above.

"I am out this evening to dinner with the Deanes of whom you have heard me talk." Del Mar's own journal tells us he went to Lincoln Grammar School in San Francisco and that one of his classmates was Louis Deane. Both Deane and Del Mar received silver medals for being first in their class at graduation. There were two eighth grade classes, and each one had a first place student. This was, by the way, an all boys school.

He writes then of some of the joys of travel—lost luggage. *"My package of blankets has gone astray. The* *express companies can rob a person here $1.25 to deliver my bag and steamer trunk. I know better than to give it to the regular express company but did not think of it."*

December 31, 1902
321 Ellis St., San Francisco, California

"I am making a little headway. A friend of mine has a mine in Mariposa County that I may soon go up to see. He is not very flush at present but in a short time expects to make a fortune. He owns a half interest in a compressed food concern. A five ounce tube contains soup, meat and cocoa and when boiled with sufficient water makes a meal that one man cannot eat. I will send you a sample sometime. It is just the thing for prospectors and army use."

"I may examine his mine for a small seam and a contingent fee."

He also reports meeting a Mr. Keely, whom he saw at Marvel and Sandy the previous year. He has no office yet but is keeping busy looking up old friends.

1903

New Year's Day, 1903
321 Ellis, San Francisco, California

Del Mar now writes that although it will be difficult, he is going to try to establish himself in San Francisco. To date he has not met with Mr. Montgomery but plans to do so the next day. *"Mining is not what is used to be here, people have been very much bitten, I will not attempt to do any business for the Eberle outfit for I would get discredited at once. Queer I have not received a word from Bisbee yet."*

He reports he went to the Tivoli and saw "Jack and the Beanstalk." Ferris Hartman was Happy Hulligan and Annie Myers, Jack.

Del Mar says he saw something new in horse carts that day *"a safety bicycle sort of buggy with shafts, that is instead of the sulky having the two wheels one on each side they were one behind the other. Rather odd sort of vehicle."*

He reports that San Francisco is experiencing miserable weather, cold and damp. He asks Belle to tell him about the New Year's parade in Pasadena (the

Tournament of Roses parade near her home). He tells her New Year's Day is a holiday in San Francisco with all the wholesale houses and big businesses closed for the day. (Apparently that was not the norm throughout the state.)

He is looking for an office to rent and writes, *"There are plenty of offices here but I find it difficult to get just what I want."*

January 2, 1903
144 Union Square Ave., San Francisco

He has a new address on Union Square Avenue. Del Mar has a friend who is the manager of the *Journal of Electricity Publishing Co.,* and now Del Mar will have his office there. Belle is to tell her father to write him at the new address if he has any news.

January 4, 1903
144 Union Square Ave., San Francisco
(A typed letter on the letterhead of *The Journal of Electricity Power and Gas*)

Del Mar reports he had met with Mr. Montgomery but can't tell yet if they will come to any terms. *"He says he may want me to go to Mexico soon and may go himself but the subject of finances has not yet been broached."* The he describes Montgomery as follows: *"He is rather an elderly man and slow of speech, but I think he understands something of mining. His Placer County property, I believe, is a good one according to his description. A company tried to work the mine but the management was bad and it did not pay. He wants me to do some testing on the area but that will depend upon what he intends to pay me for the work. I am not after anything big for, as they say, beggars cannot be choosers."*

The letter continues by saying how he misses her and telling her that she would like San Francisco. *"All dress goods here are much cheaper than in LA. I could get a lovely fur here for less price that I paid for yours. The ladies mostly wear sealskin or imitations, but the latter are very perfect so that I cannot tell the difference."*

Next he briefly describes the office where he is. *"There are three typewriters in the office here so that I can use them to write all my correspondence and what is better, I have no office rent to pay. There is nothing like having friends and especially helping one."*

"The other affair that I mentioned about the mine my friend has but has never seen may turn out something. We are first finding out from his partner who is at the property what he has and if the answer is satisfactory I will make a report. I may get but very little in cash but he is a friend that may be of great use to me."

At the end of the letter Del Mar writes this of his own father: *"My father says he may come out here and form a partnership with me if I can do anything in San Francisco. He ought to have thought of this some years ago and not let his business drop."*

The spirit in which that remark was made is unclear. Alexander Del Mar, however, was able, in an earlier time, to hold a steadier position than his son can at this time.

January 6, 1903
The Journal of Electricity Power and Gas
c/o Half Tone Co., San Francisco, California

This is a short note in which he asks *"What can be the matter that I have not heard from you since I left?"* He worries that she might be sick and asks that if that is so then for her to please have someone write and tell him.

January 6, 1903
Same as Above

This is another very short note telling Belle that he received a letter from her dated Jan. 4th but no earlier ones.

January 7, 1903
144 Union Square Ave.
c/o Half Tone Co., San Francisco, California

Del Mar begins this letter by explaining that the letter enclosed with this one is for her father and not to let anyone else open it.

Now the letter turns to business matters. *"Mr. Montgomery's affairs are not yet ripe enough to get an examination out of him but I may do a little testing for him and at the same time do some testing on a proposition that has just come to me. I was talking to a party about an iron proposition he has, some million and a half tons of black sand. He mentioned incidentally that it contained % 3.00 per ton in gold. He had never looked at it as a gold proposition and now I want to make a few tests to determine whether it can be worked at a profit. If it can there is a fortune in it providing, of course, that there is that amount of sand and it will run that value all the way through."*

This could be the break Del Mar was hoping for, and we can guess that the letter to Belle's father had some connection to this news.

January 8, 1903
Union Square
c/o the Half Tone Co., San Francisco, California

This is a response to a letter he has just received from Belle plus the ones she apparently has had returned and which she has now forwarded to him. He apologizes for giving her the wrong street name (which is what caused the problem).

Apparently in response the things she has written, Del Mar writes, *"so Mick missed his fazzer"* (father, i.e., Del Mar) and *"So Bird was in the parade and acquitted himself admirably."* Both Bird and Mick are animals belonging to the Rogers family. Bird is a horse, and possibly Mick is also. Then he adds, *"Evidently Bird likes me better than Indians for I can, or could, do almost anything with him."*

The parade to which Del Mar refers is, of course, the New Year's Day Tournament of Roses parade.

During this prolonged courtship and engagement, Belle and Del Mar have exchanged photos on several occasions. In addition, they have sent photos to his family, which is now back in New York, as well as to Belle's parents. He is glad his parents remembered Belle and says his mother sent him her own photo and will send another to Belle. He describes his mother, saying, *"She is such a dear sweet woman."* Then he jokes that he receives his disposition from her.

Since he is now in town and largely among old friends, Del Mar is having a bit of social life, which was not possible at the various mine sites. He writes that he is going to dinner at a friend's, visited the Golden Gate Park the previous Sunday, and will be dining out again the following Sunday. Despite his efforts, nothing in the way of a job has come yet.

"Mr. Montgomery is so absorbed in religion that he has little time for mundane affairs. I called Mr. M. not long ago and he asked me into 'the meeting.' I thought it was a mining meeting, but no, he has a front room on the same floor as his office where a religious meeting was in progress." Del Mar describes the meeting as, *"The usual Salvation Army style. I was asked to testify which I think was an impertinence of their part. I declined. Then,"* he continues, *"I was asked what I wanted them to pray for that I wanted. I said I did* [perhaps he meant did not] *require anything by shaking my head. They were a crowd of fanatics, groaning and calling out such phrases as these— Oh God, Bless him. Glory! Hallelujah!, etc."*

Del Mar writes that he is still trying to get financing for the sand project he mentioned earlier.

January 10, 1903
San Francisco
(Typed on the letterhead of *The Journal of Electricity Power and Gas*)

Del Mar writes that he is leaving San Francisco. *"I will leave here for Los Angeles Monday morning...I have accepted the offer of the Bisbee party although it is not pre-cisely what I want for as you notice I bind myself for a year."* (Perhaps he sent her a copy of the agreement.)

Since he has been without work for some time and funds are short he is anxious to get to work. *"I consider my visit here to be a great success for what it may bring me in the future. While I hate to give* [up] *the investigation I am making into the sand proposition yet there is nothing sure about it and there is so something tangible about getting a salary every month and I had better do this last until I get sufficient funds ahead."*

In this letter, Del Mar notes, *"The typewriter in the office* [the typist, not the machine] *knows Fred Shaw very well. I gave her some points to tease him on and she is going to have fun with him. She says they are very nice boys that Fred thinks he is a great lady killer."*

February 9, 1903
The Overland Hotel, Reno, Nevada

Del Mar has traveled to Reno on his way to Pine Grove and reports being in Reno and says he will take the train in the morning for Wabuska. *"It was so disagreeable in Sacramento that I only stayed a few hours before taking the afternoon train and arrived here about 9 p.m."*

He describes the Overland as being *"first rate although not large"* and adds that there are *"electric lights and running water in every room. Reno,"* he says, *"is quite a big town and active."*

There have been storms in California, and Del Mar describes how it has affected him. *"The snow in the Tehapichi's was about 6 inches deep and up here about 4 feet. The railroad goes through about 40 miles of snow sheds climbing up to the summit. There were people on the train who have only come up to see the snow and brought their children."*

Del Mar wonders if the storm responsible for the heavy snow hit her area. *"San Francisco,"* he says, *"had a deluge which stopped all traffic. There are fears the excessive snow will cause floods in Sacramento and the San Juaquin Valleys."*

February 1903 (no day)
Pine Grove, Esmeralda Co., Nevada

Despite the rugged winter weather, described in the following letters, Del Mar will years later write of his time with the Nevada Chief with wonderful descriptions of the area. On traveling to Yerington, he writes, *"It was necessary to pass through desert and mountain country. In no other country that I have visited could be found the contrasts to be found in the Great Desert of California and Nevada. Like the Sahara Desert, there are long stretches of sand, but generally it is a region between mountain ranges covered with sage, greasewood, and cacti. Many small streams of water issue from the mountains which flow but a little way before they sink into the desert gravel. A common saying in the desert is that you 'climb for water and dig for wood.' This refers to the water in the mountains and the roots of the sage, greasewood, and mesquite, particularly this last. The petrified or hard cactus roots make good fuel for the forge."*

In this letter, Del Mar has traveled to Pine Grove by way of Reno. He mentions having sent a letter from Reno and is now at Pine Grove. *"I will have to walk the trail tomorrow (3 to 4 miles) as the road has not been broken and is covered with snow."* Then he describes Pine Grove as *"a place of some 15 houses, mostly unoccupied. The hotel,"* he finds, *"is a very nice building for the place. Snow is everywhere but not very deep and is fast thawing."*

Del Mar is apparently in the town of Yerington for the night. His description of his time there is mostly of the people. *"Yerington,"* he writes, *"is quite a town in an agricultural country."* He compares it to *"dropping into an oasis on the desert. They mostly grow alfalfa and livestock. Hay is $5 a ton loose and not easy to get baled."*

"What a difference coming here as superintendent to what it would be had I come in a subordinate capacity. It is this and that Mr. Del Mar. Have a drink? No thank you. Why then have a cigar and once I had four cigars in my pocket and one in my mouth. The people of Yerington are farmers not miners. It is a quiet little burg."

The timber there, he reports, is like that at Black Hawk—piñon, and the rocks are like those in the desert at Gold Peak.

February 20, 1903
Pine Grove, Nevada

Del Mar is back in Pine Grove but is currently on his way to Yerington with a $1,000 bar of bullion. He is to return to Yerington by stage on Monday. Pine Grove is about twenty-five miles south of Yerington.

The weather is still flexing its muscle in the region. *"We are practically snowed in at the mine and cannot start the mill for another month and even then the scarcity of water may bother us."*

The storms he mentioned while still in California have, indeed, touched Pasadena. In response to Belle's letter, Del Mar writes, *"So you have felt snow for the first time."*

"The people at the mine were very genial. The Mrs. [no name here] first met her husband at Daggett when the King Mine was running but I do not know whether she is a Daggettite or was only visiting. She is rather nice looking and the children are cute and rather well behaved. The little girl 5 years of age dresses herself except for finishing touches and the girl 7 helps with the dishes. The little boy about 3 years old gets up at 6 o'clock when we get out and always awakens smiling. The little girl 10 years old is at school in Pine Grove." The lady he is describing, we later learn, is Mrs. Knight, who is the camp cook.

The trip back to the mine was difficult, as the snow was three feet deep. Neither he nor the young man he was with knew the trail, and besides, it was covered with snow. They went *"rather blindly but got there in 5 hours walking."* He was *"completely played out and my feet nearly frozen."*

The day he writes the letter Del Mar took the same trail but in only two hours, as now the trail had been broken by horses and men using it.

Del Mar and perhaps the Rogers family must have tried raising some calves, which have now died. Del Mar sends regrets about the calves, saying they were pretty and gentle. He asks Belle to send the strap and keep the other things.

And there is apparently time for a little fun in the camp. *"The great pastime here in Pine Grove is coasting down the hills hill on sleds, but the worst is walking up again."*

Finally, a note on Pine Grove's connection to the

rest of the world: *"The stage starts tomorrow at 5 a.m. and gets to Yerington before 11 so as to make connection with the stage to Wabuska."*

February 21, 1903
Yerington, Nevada

Del Mar is in town overnight awaiting the stage on Monday. He has sent her an Indian basket. *"I bought* [the basket] *here from an old squaw. I thought it a good piece of work. I paid $1.25 for it which I thought was cheap."*

Although Del Mar has left Johnson, the whole matter of the money Johnson owes Del Mar is apparently still unsettled, as indicated by the following lines: *"I noted what you said about R.A.J.* [Johnson]. *I can do nothing for my judgement is against him and not the mine although I can make a bluff."*

The weather is uncertain, and he hopes it will not storm before he returns to the mine.

Del Mar has come to Yerington to send bullion to the company headquarters. *"I expect the $1033 I sent the Company yesterday will make them feel a bit happy. They are behind with their accounts and it seems are depending altogether upon the mine to pay. It will be alright* [his spelling] *for a couple of months and perhaps longer, according to what is developed. It is a mine that has been coyoted for the last 30 years and is uncertain, but altogether depends upon luck and good management."* "To coyote" something apparently means to mismanage or to ravage a site without any heed to preservation of a place.

The reference to the need for good management is a theme we have often heard Del Mar repeat. Poor management seems to have been a major cause of mine failures.

Mining accidents of note quickly made the round of the other camps. Del Mar relates one incident to Belle: *"Perhaps you read of the explosion at the Douglas mine near here, one man was killed and several hurt. It was due to thawing out powder near a fire. 150 lbs. went off. One man was thrown 120 feet but not injured very bad."*

There are additional weather problems. *"I am in hopes the weather will get warm and melt the snow at the mine. At present we have only enough water to run the boiler and every pipe must be emptied each night to prevent freezing."*

March 1, 1903
Pine Grove, Esmeralda Co., Nevada

Del Mar is going to Pine Grove tomorrow and hopes there will be a letter from her waiting for him.

The letter begins by stating a variety of facts ranging from mill operation to food.

> *"We have no water at present so are not milling any ore. The snow has melted off the south side of the hills but not on the north where there is about 3 feet. I had some fresh beef packed in during the week which was quite a relief from ham and bacon."*
>
> *Mrs. Knight* [the camp cook] *does well considering she has three little ones to look after. He misses the jams and preserves at your* [Belle's] *house. "The canned jellies they have here are the poorest sort of stuff, gelatin colored with cockineal.* [He probably means cochineal, which is a form of a red dye.]

All the fruit that is in them is pasted on with the label."

Del Mar has sent the company a bill for his expenses from Los Angeles and is waiting to see what they will do with it. *"I believe this is a regular Los Angeles outfit and you know what that means."* The tone of this line suggests that *"regular Los Angeles outfit"* means "slow or never paying."

The sand project back in San Francisco is not a dead issue. *"I have heard from Dr. Deane about the sand proposition. He said he had been sick but might in a month's time have things arranged. I will be satisfied if he takes two or three months. I told him I must get $150 a month for investigating the problem."*

Sill looking to better himself, Del Mar has continued to run an ad seeking employment as a mine superintendent.

"Mr. Montgomery has answered my ad in respect to superintendant not knowing it was me. He said he wanted a man for California or Mexico. I wrote back that if he could make me a better offer than I have I would work for him."

March 8, 1903
Nevada Chief, Pine Grove, Nevada

March came in like a lion, and he and the whole force at the mine were out on the 3rd trying to break a road through to Pine Grove. *"The next day a horse and sled made it half way over but turned back and today I and another went half way and met the sled and got it through, making it the first vehicle to get in this winter."*

Del Mar thinks that now the road is broken open things will get better. He expects a 10,000 gallon tank to arrive soon and notes that he needs to get it in as soon as it arrives so he has lost no time in getting the road ready.

Food is still rather poor in the camp. He wishes he had cream, of which she speaks. *"I have none, only a tin cow,"* he laments.

The horse named Bird is again mentioned. Belle has apparently written that he is a high tone horse, and Del Mar agrees this horse doesn't belong behind a plow.

Nothing is happening there, Del Mar writes. It is just routine and snow, snow, snow. They can't start the mill because of the snow and cold weather.

> *"A miner has just arrived, walked all the way from Yerington. I have no work for him so will feed him overnight and send him on his way unless he wishes to chop wood. They usually don't care for this sort of work. I think he is the sort that travels about getting a little work here and there."*

The sled brought little more than 120 pounds of beef. They will eat porterhouse for dinner with potatoes, beans, and peas and pie—none as nice as at Belle's home. He still misses the jams and preserves at Belle's home.

March 8, 1903
Pine Grove, Nevada

Del Mar writes that he is belatedly celebrating his birthday by writing to her.

"Today," he writes, *"our forces were out over the road to Pine Grove shoveling out snow drifts so that we can get a team through. We are getting short of powder and provisions*

*and must break the road. We covered about three miles of
the six and found it drifted in places but not as bad as we
imagined. A one horse rig will try to get over on Thursday
and so make the road for teams to get in."*

He encloses a newspaper notice sent to him by Mr.
McMannen. It is a notice that Del Mar is a respected
mining engineer who has recently been appointed
superintendent of the Nevada Chief company's mines
in Nevada. *"He* [McMannen] *says the company is very
well satisfied with the progress I have made. As usual, I am
working hard, there is so much to do to get ready for the mill
when it starts running. The system of mining carried out
here is not as I think it should be so I have been changing
the method considerably. I have only four men now and
about 250 tons of ore broken and in chutes."*

Ore was usually stored in chute, which could, as
needed, be emptied into cars to carry the ore from the
mine to the mill. Here Del Mar is letting us know that
he is ready to move the ore as soon as the mill is ready
to receive it.

"So far," he continues, *"they are letting me have my
own way which I hope they will continue to do for my benefit and theirs."*

He describes life at the camp as follows:

*"We do not all live together here. There is the
boarding house with four rooms—kitchen, dining
room, and two bedrooms. Mrs. Knight occupies this.
Then there is a bunk house with sixteen bunks and
then my office—three rooms—one store room and
office, assay office, and a sleeping room with a
carpet on the floor. Mrs. K. is a very fair cook and
altogether nothing fancy yet good plain food."*

The weather still prevents running the mill. March winds have arrived. Del Mar describes the place as ideal for summer camp, like Bear Valley but with piñons instead of pines.

March 16, 1903
Nevada Chief, Pine Grove, Nevada

This letter begins with a discussion of a newspaper clipping Belle has sent. It is about a woman (called simply M. H. in the letter) whom Del Mar describes as *"a big dunce to proclaim her disgrace to the world. I have no doubt she was to blame as much as he so I have little sympathy for her. Perhaps now that she is a woman with a past the Orpheum might take her on the circuit. Of course, I look at these things from a man's point of view and as such I would advise you not to be more than friendly with her. It seems to be human nature than when a man or woman takes the downward path they try their best to take others with them."*

He has forwarded to her a book Maud (his sister) sent. Del Mar says he read it and it is *"distinctly English"* and a cute story.

The weather in Pine Grove still cold and the mill still is not running.

He writes, *"The Los Angeles* [people] *are slow about anything that concerns money. Here it is the 16th and no check. I am writing a rather stiff letter to them on the subject. I am afraid I have got myself into another 'Los Angeles Company.' I am looking for another place but will not give this up until I am sure of another."*

On an entirely different subject, Del Mar tells of reading, but not fully understanding, a paper written by his brother.

"I have read my brother's last paper but the more I read the less I understand the system of philosophy of it all. I will write him some day to know what his paper is about."

Eugene Del Mar, an older brother of Algernon, lived most of his life in New York but did move to California about ten years before his death. By the time he died in the mid-1930s, he was blind. This was, in part, due to a belief he had that looking at the sun had health benefits. Most members of the family found his books on philosophy difficult to understand.

This letter also contains lines telling how he is anxious to get to a place where she can be with him, as he misses her and her companionship a great deal.

Here Del Mar reports that the sand proposition is running slow and he has heard nothing lately. *"Maybe it is all smoke,"* he says.

Again, we see evidence that the newspaper was important in Del Mar's life. He tells Belle he *"would greatly welcome the Sunday Times."* And apparently the paper to which he mentions subscribing earlier (in order to be in a sweepstakes) wasn't a very good bargain. *"I received a notice from the Examiner that my subscription ran out on the 2nd of March and I haven't received a paper for at least six months. Someone in Victor must have been getting them."*

March 21, 1903
Pine Grove, Nevada

It seems Mother Nature is determined to keep Pine Grove blanketed in snow. Del Mar reports there was another snow storm on the 16th (after he wrote the last letter). Fortunately, warmer temperatures have arrived and the snow is melting rapidly. Once it is gone, he says, they will have the stage coming to the mine three times a week.

In response to some questions she has asked about pictures he sent her earlier he tells her the children do belong to the man in the picture, that his house is the lowest on the left, and that the trees are piñons, not pine, and very welcome.

Two new tanks he requested a month ago are on their way to the camp, and the men are preparing to run the mill. (These tanks are most likely for storing water as gold mining operations depend heavily on a steady supply of water.)

He asks her to keep the negatives of some pictures he sent. He also remarks that although he doesn't mind paying 25 cents to send her a letter that same amount would buy two ice creams and a stick of candy. It is probably safe to say Del Mar had a sweet tooth.

And he has finally received a paycheck!

"They sent me my check but the men nothing. They are all complaining and may cause trouble. Of course, I did not let them know I got anything. The treasurer sent $200 with instructions to use it where it would do the most good. It did not take me long to decide that the most good would come by paying my salary and my expenses from Los Angeles. One of the men is four months behind with his. A regular Los Angeles Company," he comments, and then adds, *"The mine is rather a hard proposition for a five stamp mill and I will try to get out of it as soon as I find a congenial place somewhere else. The Company seems pleased with all I have done but so they should for I have as usual worked hard and conscientiously."*

March 30, 1903
Pine Grove, Nevada

The weather has improved and turned to fine spring weather. The snow has melted rapidly, and water is now beginning to run in the canyons. *"This water question is the great one here and now it is beginning to come our way."*

Del Mar writes that he has sent several letters to the company regarding pay for the men, and *"the last one brought them to their senses and now everyone is paid up to the 1st of March."*

> *"We started the mill on the 21st and had run a very short time when a wall in back of the ore bin gave way and smashed the pulley that runs the rock breaker. We took a pulley off of some old machinery and now we are ready to run again, in fact was running when I left for Pine Grove."*
>
> *"Mr. McMannan has written several times that the Company is well pleased with my work and that they appreciate that I have mastered the difficulties as they came along."*

The myriad problems that Del Mar has encountered and often solved are listed in the next paragraph.

> *"What with lack of water, claim jumping, water jumping, contracts for wood and other duties I have had my hands full. The men previously in charge here were a bunch of dunces, did not locate right, exaggerated everything and I have got to try to straighten out the tangle."*

That he misses Belle and wants her nearby is obvious as the following lines detail: *"I am almost afraid to travel to Los Angeles now, too many highwaymen. Might get held up. I think you had better leave too or they might hold you up some day."* Then he reminds her, *"There is a safe place up here when things get too hot on the electric cars. Summer is coming now and it will be fine up here and no highwaymen and I have a house with three rooms."*

The warmer weather they are having means improved contact with the outside world. *"We will soon have the stage running in here* [since] *the roads are thawing. A buckboard and one horse came over with meat, eggs, etc. Beef,"* he notes, *"is 8½ cents a pound and eggs are 20 cents a dozen in Yerington."*

The mention of the cost of eggs must have prompted Del Mar to ask how her chickens are doing. He makes inquiries about Deb, who apparently chewed on Belle's shawl, and Mick, who has a habit of running off. Both Deb and Mick are chickens.

April 6, 1903
Pine Grove, Nevada

Del Mar begins by saying he has received the balance of photos she sent and a telegram she must have also sent. The telegram was from the Golden Link, as Del Mar writes, *"I do not know what the Golden Link wants and will await a letter from him. You did right in opening the telegram and answering it."*

A hint that this long engagement might soon end comes as Del Mar writes, *"I go to sleep thinking of you very often so you ought to dream of me a great deal. Perhaps the dream will come true soon—one can't tell."*

The Pine Grove area is still plagued by winter weather as they again have bitter cold weather following yet another snow storm. With a bit of wry humor, Del Mar says, *"A thermometer would be ashamed to register up here so we have none and consequently we do not know how cold it is except by feel."*

April 10, 1903
Pine Grove, Nevada

Earlier letters have made reference to someone named Fred. He is someone Belle knows who frequently gives her pictures of himself. Now Del Mar responds to something Belle has said about Fred. *"Fred is giving you a fine josh about the pet he most longs to care for."*

"I have sent in my resignation to the Company unless they straighten out the business I anticipated while I was in Los Angeles. Mr. Knight says he was employed irrespective of any manager here so I decided to leave unless the Company upheld me." Mr. Knight is the husband of the lady who has been cooking at the camp. He was the millman and apparently felt Del Mar's position was not superior to his own.

"Mr. Montgomery [of San Francisco] *has offered me a position on the 12th of June at $150 a month but I have to run the mill part of the time as well as superintend so even if I give this up I will not be long idle."*

Travel to and from the mine is improving as the weather improves. *"The mailman came over today and will make regular trip twice a week in the future. The first freight team came yesterday and sprung both axles in a snow bank which nearly tipped the wagon over."*

April 12, 1903
Nevada Chief, Pine Grove

In a lighthearted note, Del Mar says he has found a lady to help Mrs. Knight and pays her $5 per week *"so see what you might have gained had you been here."* He continues describing the new woman, saying, *"She is smaller than you are and is the wife of the foreman."* The statements which immediately follow tell us a lot about the economics of running a mill. *"I anticipate the Company will side with me and out will go the Knight family. I have nothing personal against them but I can save the Company over $3 a day or 67 cents a ton of ore milled which is considerable. As it is now, the Company pays him $5 a day and feeds his whole family (four children) and Mrs. Knight has to have a helper at $5 a week and her feed while I can get a millman for $4 a day less $1.00 board and a cook for $40 a month."*

At the end he laments, *"If we only were not so far apart how nice it would be for you all to spend the summer here in the Mountains. Let us hope at any rate that better days are coming."*

April 19, 1903
Nevada Chief, Pine Grove, Nevada

He has received an answer to his letter of possible resignation, and it is satisfactory. The Company gives him full and absolute charge of the site. The Knight incident is closed, and the entire Knight family is now gone. Del Mar is running the mill until he gets a new millman, which he thinks will be soon.

"I have a better chance now of making a success of the property for the Knight family were quite an expense to the Company and he was not a particularly good millman, was careless and wasteful. I think I will make a success of the enterprise. Anyway, I am trying hard."

The weather is getting very nice and springlike. He still wishes she and her family could spend the summer there. He hopes to visit her in August or September.

Del Mar now mentions the plans he hope will come true for him and Belle. *"I think before school opens again we had better join hearts. This would allow your mother to stay home and you could go home at Christmas and stay two or three months until the winter is over."*

"It is no use putting it [marriage] *off forever or we will never get the opportunity. Of course I expect to get better places than this one here. This experience will be the most valuable I have ever had and I have no fear now of being able always to find a position. I have more confidence in myself and can tackle anything from running the mill to the underground work."*

Please remember that what has delayed the marriage for so long is that Del Mar wants to be in a secure job with a decent salary. His confidence about being able to do just that seems to have grown tremendously by his experience in Nevada.

He tells of looking for a cook. Currently the foreman's wife is doing that job. He is offering $40 a month, and most applicants want $50. *"The wages in Nevada are higher than in California."*

April 27, 1903
Nevada Chief, Pine Grove, Nevada

Del Mar speaks here of being rushed but also reports the new millman has arrived and he will have more leisure now after having to run the mill himself the previous ten days. Things must be going very well, for now he writes of hiring another millman. This will allow him to have the mill running night and day as long as the water lasts. *"The company has raised my pay $25 a month without my saying a word and promises another raise in the near future so your honey is making things move about here and is appreciated in Los Angeles."*

He tells Belle his mother would like to hear from her and comments his mother is very busy of late and behind in her own correspondence.

May 4, 1903
Nevada Chief, Pine Grove, Nevada

He has received her letter, in which Belle has accepted his proposal regarding the time and place for their marriage and reported it to her mother. *"And now that you signify your willingness to cast your lot with mine when I appear in Los Angeles to get you. I enclose a check for $12 which you can use to get your photo taken or in any other way you desire. This is a birthday present delayed in transit."*

He also asks her to send him some photo supplies connected—a tube of Luxo Flash light powder, a roll of films (6) 5x4, a package of soho paper—24 sheets 4x5, and one ounce of sulphocyanide of potassium.

He agrees that it would be nice to go to the beach

for a couple of months but reminds her that it would be hard for him to get away to visit her there. He does promise to look into the possibility of finding enough time to go there.

The previous letter from Del Mar seems to indicate the marriage will take place in Pasadena. Belle, however, must have written some other suggestion to prompt Del Mar to write, *"I suppose you would care to go to San Francisco with say Miss Stevens and get married there. I think she will go visit Mrs. Shawhan this summer."*

Perhaps in an effort to make the mill camp seems an inviting place to live, Del Mar writes, *"We are likely to have two women here cooking so there would be company. Mrs. Fish who is at present is cooking is a young woman, a splendid cook, and a fine woman in the bargain. She is the foreman's wife, unassuming, clean, and respected by everyone."*

The mill is producing, as the following tells us.

"I sent the Company a bar of bullion worth about $2000 for the April cleanup. They ought to give me another raise in salary soon. I expect this month to send them $4000 or over. The mine is looking well, I have their confidence, and I think I can make a success till the winter comes again by which time more good ore might be developed. Everyone who has ever worked this mine made a failure of it so I want to be the exception."

Belle has written again of the horse Bird, which she reports as being quite frisky. Del Mar says that the frisky behavior shows he is happy and well fed.

And still speaking of horses, he writes, *"I expect to*

*get a team next week for the Company. Two grays, one a
good single buggy horse. I can save the Company the cost of
the outfit (about $300.) in three or four months."*

Del Mar is probably anticipating being able to go for
parts and supplies without hiring a wagon to do such
jobs. In that way, the new outfit would pay for itself.

May 11, 1903
Nevada Chief Mine, Pine Grove, Nevada

Dell Mar sends clippings from the Yerington paper,
where, he says, *"We are referred to as Rockland and has
been known as such for years."* According to the article
they had a visit from the Pine Grove Ladies.
*"Unfortunately I was busy at the mill all that day so could
not show them any attention. The schoolmam* [his
spelling] *at Pine Grove is quite an attractive young lady
and one of our men Perry Morgan is greatly smitten and
comes in for a lot of joshing."*

There seems to be some unrest among the miners
regarding wages. Del Mar *writes, "Four of the mines struck
for higher wage and got their time. They were not satisfied
with $3 a day and eight hours so I let them go. As I always
have from 150 to 200 tons of ore on hand I can do without
miners for a month if necessary."*

On a lighter note, Del Mar writes that he believes
that somewhere in Los Angeles, maybe in the window
at Montgomery's, Belle can see the last two bars of gold
bullion he took out for the Company.

The water problems do not go away. Del Mar now
writes that the water supply is getting short and they
will soon have to pipe water from a spring 1¼ miles
away.

Good weather has come to the area, and Del Mar wishes Belle's father were around. *"If your father were only here I could make a deal with him. I am trying to buy a team of horses about 1300 or 1400 lbs. with wagon and harness for $300 but have found none to suit. Horses are pretty high about here this year although feed is only $7.00 a ton."*

He took bullion to Yerington on May 2 and returned the same day on horse, *"making in all about 56 miles by stage and horse and even then I got to the mine in time for supper at 6 p.m."* He expects to have the machinery covered as early as the next month.

In closing, he admonishes Belle, *"Don't talk of strawberries. You make my mouth water."*

May 18, 1903
Nevada Chief, Pine Grove, Nevada

Belle has reported hearing and seeing Teddy Roosevelt, and Del Mar wants to know how he compared with her impression of McKinley.

Despite the nice weather reported in the last letter, now Del Mar now tells us that there was a big snow storm the previous day. The eight inch snowfall forced them to close the mill. *"It is bitter cold for this time of year."*

Horses are still a current topic for Del Mar. *"I had hired a team instead of buying one. It is a fine pair of bays, young and high lifed, one six and the other the magic eight. Small feeders, big workers, high life."*

"The Los Angeles people are so firmly impressed that they have the only mine in the world that a $10,000

bar would not surprise them. They expect $6000 this month and it is all I can do to get them to expect less. Water was getting very short until the snow came. We may have enough till the end of the month."

Still missing Belle and wishing very much she were there, Del Mar writes, *"If you come up here with me we would have a house to ourselves and would board at the boarding house or keep house as you pleased. I think the former preferable. I think I could get a piano but not a very good one so that you could still keep up your music. I would do all possible to make you happy and contented while here and then about November or December you could return to Pasadena for the winter or I might be in some more pleasant locality."*

Here Del Mar gives us more details about the camp and the people in it. *"We have two lady cooks now, both married and with their husbands working here so in case one goes we will still have one on hand. Our whole place is about 17 so you see we have quite a family to feed and have to divide them into first and second table. We have fresh meat all the time and had some fresh milk but the last lot spoiled on the road, it being too hot. Can't get chickens less than 75 cents a piece about here so we have not had any chicken dinners."*

May 25, 1903
Nevada Chief Mine, Pine Grove, Nevada

Still no let up in the bad weather as Del Mar writes that an additional 18 inches of snow fell on Saturday. *"We have hardly yet had spring here but have hopes,"* he writes.

He tells Belle that if she wants to see some gold bars she should come there around the first of the month. The first of a month is apparently when Del Mar shipped the bars to the company.

Once again there is also talk of photos as he thanks her for sending the Kodak supplies and reminds her to send two copies of her own photo so he can send one home to his family in New York.

"The mill keeps right on," he says. *"We even had it running while the snow was falling but it got too cold for the amalgamator.* [This is the plate where the gold is combined with the mercury before the gold is finally separated out.] *I think I will send from 3 to $4000 to Los Angeles this month. There is week yet* [in May] *and if everything goes well we will run 24 hours to make up for lost time."*

The mill is not without visitors. *"Two pretty girls were out during the week. They came to visit Mrs. Fish, but more particularly, I think, to see some of the boys employed here. I know one* [of the men] *is stuck on the younger girl. We have here four or five boys who live on the River and are consequently well acquainted besides which all the people in the Valley seem to be more or less related to one another. The beau of the schoolmistress at Pine Grove is employed here and comes in for a lot of joshing."*

He closes the letter after noting there are twenty-two men working there now and characterizes them as *"Quite a gang and something to provide for."*

June 2, 1903
Yerington, Nevada

Here Del Mar explains he is in Yerington with a $2,606 bar of bullion to ship to Los Angeles. He notes that one of the directors, a Mr. Garretson, and his wife may be visiting next month at the mill.

That George Rogers is always on the lookout for a big strike is evident again in this letter. Del Mar comments, *"So your father is after the lost Lee Mine—I hope he has luck! Write him that if he gets down on his luck and is within driving distance of here that I will hire a team from him and feed it and keep the wagon in repair for $1 a day and pay him regular wages $3 day less $1 board for driving it."*

As for the new team that Del Mar acquired not long before this, he writes, *"I have only ridden on the team once. I have a teamster to load wood and haul it to the mill and drive to town when necessary."*

Belle must have sent a strawberry recipe despite Del Mar's earlier admonishment to not speak to him of strawberries. He responds, saying, *"Well, I guess I do wish I had some of your strawberries and what a fine receipt. To supply our camp with a dish like that would cost a fortune."*

Here Del Mar writes that the piano he mentioned earlier is at Pine Grove, and *"I could hire it for a song as the lady is away nursing at a hospital."* There is no indication as to why the piano is mentioned, but perhaps Belle's music lessons had included the piano and Del Mar thought she might like to have a piano at or near the camp.

Now some more on the wedding plans:

"I never thought for a moment you would consent to go to San Francisco to meet me for the reason that I thought your mother would want to be present at our marriage but you talk it over with her and if she does not object perhaps we can arrange the matter. I see

154

*no harm at all for on your arrival in San Francisco.
You could stay with one of my lady friends and the
next day we could be married. It will be a month or
two yet and lots may happen in that time so let's see
how circumstances will govern our case. If we keep
putting it off, we will both be old maids. I know the
daughter of a bishop who went all the way from
London to the center of Africa to marry the man of
her choice. She was a small precious parcel like you
and had plenty of grit."*

*"The country here is looking fine, everything is green
and the alfalfa almost ready to cut. I had to pay $7
a ton for it but next month it will drop to about $3
or $4 per ton."*

June 8, 1903
Nevada Chief, Pine Grove, Nevada

Just when the wedding seemed around the corner,
more bad luck appears. *"The news I have to tell you this
week is that I may leave here sooner than I thought. Our ore
body has all of a sudden given out and unless we strike other
ore, the mill will have to shut down about the last of July.
You may see how uncertain mining enterprises are. What
the company intends to do I have yet to find out. One of the
Directors will be here in July but I am asking him to come
sooner for it is difficult to convey in a hundred letters what
half an hour at the mine would do."*

He has already written Mr. Montgomery to see what
he has to offer and he has put his ad in the mining paper.
*"It may be that we will strike ore at any time and keep right
on but it looks rather blue at present. I hope that if I do change
places the next will be close to San Francisco or Los Angeles."*

June 16, 1903
Nevada Chief Mine, Pine Grove, Nevada

Fortunately the news now is better. *"We are having fine weather. The mine looks better than last reported, having struck ore in two places. Perhaps things are on the turn for the better."*

"So you went in the water at the Beach. I suppose you had a black bathing dress, one of those baggy affairs." No further word appears on that adventure.

Big plans are in the making for a Fourth of July celebration at the camp. *"We are to have a chicken dinner with ice cream for the Fourth of July and will close the mill for two days, clean up a bit, and make repairs. If your father would only come here with a good team, he could do some trading in Mason and Smith Valleys and it would be easier for you to come here."*

Again, Del Mar tells Belle that the last gold bars are on display *"somewhere in Los Angeles. Next time you go in* [to Los Angeles] *stop at the different jewelers' windows."*

And then it is back to wedding plans:

"If all goes well, let's set our date for about Sept. 1. I will know in a month's time how things will go here and that will be in time for you to make preparations and if things keep going well we will visit St. Louis next year, so there now. If I can get to Los Angeles, I would certainly prefer to do so, or if not, why could not your mother come to San Francisco with you and return when we leave for here. If this arrangement is carried out I will insist upon standing her expenses or it is no go. She could leave the kids for a few days with your grandma."

Water and getting a good supply of water is still a main concern at the mine. Remember that earlier Del Mar talked of possibly needing to pipe in water from slightly over a mile away. In this letter he says, *"We have laid a pipe line about 9000 feet long from a spring the Company owns but other parties claim. They* [the other party] *have pulled the pipe out so a lawsuit is in view. More food for the lawyers."*

From Del Mar's work diary we learn that at some point two prospectors located a placer claim that covered the Nevada Chief's water supply. Del Mar went to the county seat, Hawthorne, and hired a local attorney; the two prospectors hired a San Francisco lawyer. Del Mar defends hiring the local man by saying that the local attorney usually has more sway with the local judge.

Unfortunately, there were numerous delays before the matter was settled. Del Mar references those delays and problems in his letters.

June 20
1903 Hawthorne, Nevada

Now begins the frustrating series of letters regarding legal action over the water line. *"I am in Hawthorne, the County Seat on law business in connection with parties who tore up our pipe line and threatened to blow it up with explosives if relaid. I am having them arrested and bound over to keep the peace and may be here several days."*

He describes Hawthorne as being much like Victor *"out on the desert but near Walker Lake. This lake lies right in the hollow of the desert, is some 20 miles long and three or four miles wide but it is all desert about."*

157

June 27, 1903
Hawthorne, Nevada

Del Mar writes that he is still in Hawthorne awaiting the results of the lawsuit. He has been there a week and is *"thoroughly sick and tired of the place."* The change in climate and water caused him to be quite sick, and he had to call a doctor. It was nothing serious, he says, but reminds Belle that his work requires him to be in good condition.

The hearing on the lawsuit has started. *"I was on the witness stand for over six hours yesterday. I think we will gain our point and will probably finish tomorrow and this will give me time to get back to the mine by the 1st of July."*

He adds that he has enclosed a letter to her father and asks her to forward it to him wherever he is. *"I wish he would come up here,"* he repeats.

"The mine is improving and looks considerably better. We have a three foot ledge in one drift just opened." Then he adds that the gold bars have been sent to the mint this time. This is important because money for the gold comes from the mint. *"I think I will demand from the company a week or ten days vacation about the 1st of September, take the train to Los Angeles, claim the girl whom I love with all my heart and soul and return here for a time and probably your father will be here so all would be pleasant and well. We will call the date a Maybe and let us hope it will be a Sure thing."*

"You know we have a date for St. Louis so even if I have to pawn my new $1 watch we will make the attempt. How nice it is to build castles and how disagreeable to have them shattered."

St. Louis was evidently the place the couple would like to visit for their wedding trip.

July 5, 1903
Nevada Chief MTM Co., Pine Grove, Nevada

The good news is that the mine is prospering again. Del Mar writes, *"The mine is looking first rate now in fact better than it ever did. I am shipping $2600 for this month's run and expect to get out $5000 for July if things are right. The parties whom I was trying to have keep the peace for disturbing our pipe line were discharged. There is no doubt the water belongs to us but this question did not come up in the present action. The Justice of Peace before whom this was tried must be a big fool for our evidence was fully verified by the admission of the defendants."*

Del Mar repeats that he hopes her father will accept his offer to come to Nevada and feels he might be able to do even better now than he had stated in an earlier letter.

"I returned from Hawthorne on Wednesday and was glad to get back to work again. Most of our outfit have gone away for the 4th, only three and myself remaining and I am again doing the cooking until tomorrow when the cook returns."

There is no mention about the 4th of July party that was planned earlier. Perhaps the two day holiday ended that idea as the men seemed to have gone elsewhere.

"You had better hurry up and join me here for one of the finest looking girls in the country is coming here to help Mrs. Fish with the cooking. Her name is Miss

Clara Morgan so you had better look out and be ready the first week in September."

"Ask Perry [Perry Hulbert is a relative of Belle's] *whether at he might like to come to the mill and run it on an eight hour shift. I could break him into it. He would be paid $2 a day clear the first month and $2.50 a day clear after that. I am about to put up a cyanide plant which I might give him charge of. If he wishes to come I could always find something for him to do and the experience that would be valuable to him. I would not say anything about this,"* he tells Belle, *"but that I think I see my way ahead for some months."* Then back to Perry: *"Tell him to bring his bedding."*

Although it its cool and pleasant in the mountains he reports it being different in the valley where the ranchers are busy cutting hay. The hay will cost $5 a ton plus a dollar a ton to have it delivered to the mine.

July 13, 1903
Pine Grove, Nevada

Here Del Mar reacts to Belle's letter in which she apparently tells him of good times happening where she is. He regrets that duty calls him to be elsewhere but adds, *"in a couple of months you can visit me and get a breath of mountain air. It will be rather cool of evenings but a house with a stove will make it rather cheerful."*

Then, perhaps to warn her of what housekeeping will be like with him in the picture, Del Mar writes, *"I wish you could see my room. It is a fright. Clothes scattered all over, papers everywhere and things generally as if I had just left in a hurry. No fear of socks to mend for I have washed until too full of holes to mend and then they go in the scrap box."*

He is on his way to Hawthorne for more legal work about the water problem and he dreads the trip. *"Our opponents persist in turning our water away from our pipes so we will have to go right at them and have it decided to whom the water belongs. We had just begun to run the mill 24 hrs and I had put on the full force of men when they began their dirty work. This is not the only water we have but is what we require to run full time on."*

When Del Mar was last in Los Angeles he bought a watch, but it appears he has since complained to Belle that he no longer has it. She, logically, assumes the watch has been lost. Now we learn the real story. *"The watch I bought in Los Angeles, I didn't lose but tried some blacksmithing on it and put it out of tune."*

In the event that her father has not left yet, Del Mar sends directions on how to reach his location. *"I think his road is from Mohave to Keebler to Lodaville to Hawthorne and from there he could get to Rockland by a short cut through Wright's Ranch. It would be quite a coincidence should I meet him at Hawthorne while the trial was on. It must be a hard desert road but he is used to that. I don't expect him for a month or more."*

July 16, 1903
Hawthorne, Nevada

Thanks to the lawsuit, Del Mar is now back in the county seat and expects to be there for a day or two. *"I have no doubt that we will win our case and get the water that is in dispute. Our opponents destroyed our dam and drained the water from our pipe line thus leaving us short of*

161

water. *We will bring suit to quit title and serve an injunction to restrain them from interfering with the water."*

In the letter he encloses *"a small heart which I bought in a store here."* At some time Belle must have expressed a desire to have a set of silver buttons, and he said he would buy them for her. Now he writes, *"About the silver buttons, you buy them and tell me what they cost for I will keep my promise but unfortunately I can't get them here."*

There are more frustrations with the legal problems. Now Del Mar reports he could be done with the matter except their lawyer is now out of town, a fact which has prevented Del Mar from leaving Hawthorne. Had the lawyer been there, Del Mar says, he could have returned to the mine that same day.

At the end he asks her father's location and adds, *"If Perry comes I can do better for him than I stated when we got everything in running order and are doing well."*

July 27, 1902
Nevada Chief, Pine Grove, Nevada

Del Mar writes that he is glad that she is back in Pasadena and not drowned as her letter said. She has apparently been in Newport. Belle has reported she had gained some weight. He answers, *"While you are gaining weight, I am losing. You will remember I left Pasadena weighing 152, now I only balance 137 lbs. and there I remain."*

The trial is making future planning difficult:

"The date that the case comes for trial I do not know yet and my and our plans will have to conform to this. If the trial is to take place after the 1st

September some time I would like to run down about that time and carry you off and see the officials at the same time. I think this about the programme but will know better in a week or so and will let you know. You appreciate no doubt the reason of this uncertainty which is no fault of mine. As far as I am concerned, love, I want to see you and be with you as soon as possible for it is a long while to be away from you."

It is probably a good thing that a large, elaborate wedding was not being planned. A constantly unclear date would make setting reservations for the ceremony and reception very difficult.

More now on the still unidentified Terley. *"I will do as you say for you know best. My principle reason for getting him here was so that you would not feel as lonely up here but perhaps with your father here and your own true love to cheer you it will not be so bad. I will try to take your mother's place in nursing you should you get sick."*

"Your plan for Sept. 1 is a splendid one and I will try to make it come that way if possible. You are right, I need a little Mrs. D. to keep me tidy and to keep my room in order."

It sounds here as though the wedding is set for September 1. Then he adds the following about their future finances and obligations as he sees them.

"Before we start out on our matrimonial journey I want to tell you that I am a poor man financially and only have what I earn but with God's help I hope to always have enough to keep us both in good circumstances. I have been sending my mother $20 every month for some time and as we are to form a

partnership in love and everything else we must try to help both our families when in need so long as we keep sufficient for rainy weather. With the $125 I now get we can put away half for a few months to come and I expect to get my salary raised at any time."

Again he asks Belle to get her father started off or he will never get there. *"I may be able to keep a saddle horse here, that is, one of our men who has a horse may keep him if he pays half the feed."* It isn't clear if this saddle horse would be Del Mar's or Belle's father's horse.

"Yesterday I bought over $10 worth of fresh vegetables beets, onions, turnips, carrots, radishes, potatoes (new) and two boxes of apples. They are grown on the river and delivered here. So with fresh meat and chickens occasionally we are doing pretty well."

August 3, 1903
Nevada Chief, Pine Grove, Nevada

"I have written the company by this mail asking for a leave of absence to go to Los Angeles and if I obtain it and it does not conflict with our law suit I may leave to get there about the 24th or 25th of this month so that I can be back in time to clean up on the first of September or if our case comes later I may arrive in Los Angeles about the 3rd of September." It sounds as though the wedding is getting closer.

There has still been no word from George Rogers. Belle has written that he has with him Bird, the horse about which she has previously written. Del Mar is doubtful that Bird is heavy enough for the work for which he needs a horse, but if he isn't they will keep him there to use. The exact use to which he might be put isn't told. However, Belle has earlier expressed a fondness for this horse, so perhaps Del Mar will find a way to keep him there but away from heavy work just to please her.

"Our young lady cook has left to go camping and we expect another one and if the report is true she is a peach and has a hundred cattle to go with her. You had better be ready for me for with all these charming ladies about and the (possibly) more charming cattle it would be unsafe to allow me to remain an eligible unmarried man."
He compares the warm days and cool nights they are now experiencing to Bear Valley weather.

"Everyone that comes into my den of an office remarks upon the bachelor like appearance it presents and remarks that it was about time I had a housekeeper. I only occupy one room but it is a peach."

The water problem has hurt the bottom line for the mill. *"On account of lack of water, our cleanup* [amount of gold recovered] *this month fell below our expenses, the first time this has happened since I came here. I am expecting the sheriff any day to come with an injunction against the parties who are holding our water from us but things are very slow in this country and distances so far apart."*

August 9, 1903
Nevada Chief Mine

There has finally been some action to correct the water problem:

"We had an injunction served on the parties who were disturbing our water and now have enough to run full time. The tailing pond dam broke twice today and I have been on the jump all day and practically since 10:30 last night when it broke and the millman awakened me out of a sound sleep. No wonder I am only a shadow, but I feel first rate."

The trial was finally ended when the attorney Del Mar hired produced a Colorado decision that ruled that a quartz claim (such as the Nevada Chief) which covered a placer location overrules the claims of the placer claim to some degree. That ruling stated that the placer mine can only claim 25 feet on each side of his center line, not the 300 feet usually allowed. This meant that the claim of the prospectors did not cover the spring from which the Nevada Chief drew water.

"My plans at present await the answer I may get from Los Angeles to my letter asking for a week off. I will telegraph you particulars as soon as I know so that you will not have to rush. I will only be in Pasadena a day or two at the most, then a day or two in San Francisco and then to our shack at Rockland. Our present teamster, who by the way once ate four ducks at one meal and then asked if they had any more quail, was just now at the office and asked me when I last cleaned out my room. I

told him that a man who is accustomed to getting up any old time in the middle of the night had no time for sweeping out his room."

"Our mill is about 200 feet from the house but does not disturb one unless it stops running and then it wakes one up. The music of a running mill is certainly very refreshing in a mining camp and this and another are the only ones running in Esmeralda County."

Here the letters end. Del Mar must have had his answer on the time off almost immediately after he wrote the above letter on August 9, 1903, for the couple was finally married in Pasadena on August 20, 1903.

There is one final letter to the story. This one written by Belle to her family a few days after the wedding.

August 23, 1903
Occidental Hotel, San Francisco, California

Dear Mam and Children,
"This is my first letter and must be with pencil as there is no ink in this room."
"Here I am in this big city and having a splendid good time."
"I shall begin where I left off. I hated awfully [I] had to leave you all but I vowed I would be brave and I think I was a captain thro' it all. We took the car to Los Angeles and went to the West Minister Hotel and next morning were up bright and early to catch the 8 o'clock train here. I saw the Pasadena cars going home and I thought how long it would be

*before I took one home again. The train we came up
here on was a fine one. We came Pullman Parlour
Car and dined in the dining car. My they do serve
things nicely. I had my breakfast just as we pulled
out of Los Angeles."*

*"I saw so many pretty things on the way up Point
Conception and oh so much but I will write you
more exactly next time when I have more time."*

*"We arrived in S.F. at 12:30 p.m. and came here.
Talk about rice. Gee Alg was just loaded down. All
his pockets and valise so there was rice everywhere
but we did not care a bit."*

*"It is cold up here. You will be surprised to know that
I wore my tailor suit all day and hugged my jacket
and will wear my fur today. I shall not have on my
summer clothes at all. My clothes are lovely and I
have not seen a sweller suit. This is a big, big place
and cobble stones and hills. Pasadena's ahead to
me."*

*"We went to see Love's Dream yesterday and to the
chutes [?].We are going to see Mrs. Shawkare* [pos-
sible an actress performing then in the city] *this
evening and across the bay this morning.*

*"Alg says when people see my clothes they are so
stylish and probably people will think he has married a rich girl. No work but my dear mother's work.
Alg has been as good as gold and as proud as a peacock of me. We have gotten along just lovely no
scraps yet at all. Went shopping yesterday in the
Emporium. Well I must close and only send a scrap
[?] letter this time as I must go to breakfast."*

"How I should love to talk to you all but am over

500 miles from home. I should love to see you but yet am glad I am with Alg for I am happy and having a fine time here. I sent a telegram yesterday."
"Love to all my home ones."
"Your loving Belle"
"I will write again and tell all the details of my trip and sightseeing."
"Write to Pine Grove
Esmeralda Co., Nevada"

Here the letters end.

Epilogue

The young couple set up housekeeping in a one-room cabin at the Pine Grove mill site. They took their meals in the mess hall with the rest of the camp. They had three sons over the next seventeen years—Roger, Bruce, and Walter. While Belle moved with Del Mar as much as possible, she often spent winter months with her parents so that the boys could attend school. In 1914 the Del Mars built a home on Alpha Street in South Pasadena, across the street from the home occupied then by George and Rosa Rogers. Later Reine Rogers Wride and her husband, Homer, built a home next door to the Rogers' home. Belle and the children joined Del Mar on weekends and vacations, as travel would permit. By 1920, Del Mar was working again at the Black Hawk Mine and continued there for twenty years. Belle and the children spent much of their time away from school at the mine with Del Mar.